CLASSICS IN THE HISTORY AND DEVELOPMENT OF ECONOMICS

General Editor: Michio Morishima, *Emeritus Professor of Economics, London School of Economics*

In the postwar years the discipline of economics has become highly advanced by focusing upon issues which can be expressed in mathematical terms and ignoring issues upon which it is difficult to make axiomatic analysis. This series aims to make available in English texts which might well have played a major role in the development of a more balanced – not exclusively mathematical – economic theory but for the fact that they were written in a language other than English. However, the series' interest will also embrace mathematical and English-language works where these appear to have been unduly neglected. The series will also seek to make available in English important works that present the experiences of non-English-speaking economies; it is hoped that these will contribute greatly to making economics more comprehensive and more widely applicable to a range of world economies in the future.

*Titles include*:

Marco Fanno
A CONTRIBUTION TO THE THEORY OF SUPPLY AT JOINT COST
THE MONEY MARKET

Hiroshi Hazama
THE HISTORY OF LABOUR MANAGEMENT IN JAPAN

Alfonso de Pietro-Tonelli and Georges H. Bousquet
VILFREDO PARETO: Neoclassical Synthesis of Economics and Sociology

Joseph A. Schumpeter and Yasuma Takata
POWER OR PURE ECONOMICS?

Yasuma Takata
POWER THEORY OF ECONOMICS

Giulio La Volpe
STUDIES ON THE THEORY OF GENERAL DYNAMIC ECONOMIC EQUILIBRIUM

**Classics in the History and Development of Economics**
**Series Standing Order ISBN 0–333–71466–0**
(*outside North America only*)

You can receive future titles in this series as they are published by placing a standing order. Please contact your bookseller or, in case of difficulty, write to us at the address below with your name and address, the title of the series and the ISBN quoted above.

Customer Services Department, Macmillan Distribution Ltd
Houndmills, Basingstoke, Hampshire RG21 6XS, England

# A Contribution to the Theory of Supply at Joint Cost

Marco Fanno

Translated by Cyprian Blamires

Forewords by
Lionello F. Punzo and Michio Morishima

in association with the
ISTITUTO ITALIANO PER GLI STUDI FILOSOFICI
NAPLES, ITALY

First published in Great Britain 1999 by
**MACMILLAN PRESS LTD**
Houndmills, Basingstoke, Hampshire RG21 6XS and London
Companies and representatives throughout the world

A catalogue record for this book is available from the British Library.

ISBN 0–333–74595–7

---

First published in the United States of America 1999 by
**ST. MARTIN'S PRESS, INC.,**
Scholarly and Reference Division,
175 Fifth Avenue, New York, N.Y. 10010

ISBN 0–312–21698–X

Library of Congress Cataloging-in-Publication Data
Fanno, Marco.
[Contributo alla teoria dell'offerta a costi congiunti. English]
A contribution to the theory of supply at joint cost / Marco Fanno
; translated by Cyprian Blamires.
p. cm. — (Classics in the history and development of
economics)
Includes bibliographical references and index.
ISBN 0–312–21698–X (cloth)
1. Supply and demand. 2. Costs, Industrial. 3. Business cycles.
I. Title. II. Series.
HB801.F3613 1998
338.514—dc21 98–21472
CIP

---



First published as a supplement to the *Giornale degli Economisti e Rivista di Statistica*, entitled 'Contributo alla teoria dell'offerta a costi congiunti' by Marco Fanno in 1974



This book is printed on paper suitable for recycling and made from fully managed and sustained forest sources.

10 9 8 7 6 5 4 3 2 1
09 07 06 05 04 03 02 01 00 99

Printed and bound in Great Britain by Antony Rowe Ltd, Chippenham, Wiltshire

# Contents

# General Editor's Introduction

This series, with its designation 'history at development of economics' has at least four areas of focus, though it would be too restrictive to call them aims.

Since the last war economics has become 'mathematicised' to what could be deemed an excessive degree, so much so that mathematical models are incorporated into the analysis even of questions where there is no need for mathematical argument. As a result, those issues which cannot be expressed in mathematical terms have been all but forgotten. Moreover it has become almost impossible to establish links between economics and other social sciences, in which mathematics are little used. This increasing use of mathematics has thus meant that economics has become isolated; the isolation has in its turn promoted mathematical inbreeding.

The net result is that the discipline of economics has lost many of the capabilities which it formerly possessed. Moreover, since such capabilities have been dispensed with in the selection of specialists, it has become more and more difficult to shift economics away from the path along which it is now proceeding.

One effective means of correcting this tendency, and of giving the contents of economics a better balance, is to dig out some of the economics of the past, and to learn again from those who have gone before us. In the earlier decades of this century economics was not the overwhelmingly English-language-dominated discipline that it has become in the postwar period. There were top-class economics achievements in French, German, Italian and other languages as well. My intention, therefore, is to select from the papers and books written in other languages some which I consider to contain useful knowledge and suggestions, which may help to promote a more balanced economic theory. By translating these works into English, they will be made available to all. This is the first point we will take account of. Secondly, I will try to annex to the series wherever possible critical biographies of scholars active in a wide variety of fields, apart from mathematical economics, in order better to learn from them.

This series is not, however, necessarily 'anti-mathematical'. It is also the intention to include in the series works which might well have played a major role in the mainstream development of economics in the postwar years but the misfortune that they were written in a language other than English has caused them to remain unknown. This, therefore, is the third point we have in mind.

Finally, modern national economies have not all evolved in an identical fashion. In Japan, for example, and in some other non-English-speaking economies, there have developed perfectly viable and, indeed, efficient economic systems. Work to clarify the structures of these kinds of economic system has been accumulating, but mostly in the language of the country itself. I am also anxious, therefore, to incorporate into the series translations of works in this area, and also research on the historical experience of these economies.

Given the four areas of focus which I have outlined above, the series as a whole will inevitably have a somewhat 'motley' character. While the works may be somewhat disparate, though, I want to build up a series in which all the volumes will prove enjoyable and interesting to read. The endeavour involves a great deal of translation work, meaning that publication at regular intervals is likely to be impossible. In addition, there are many candidates whose work must be considered for inclusion. This, of course, makes a great deal of work for a single editor, and therefore it will take time for him to put the project into orbit. My fervent hope, however, is that this series, which, among other things, expresses my own philosophy on the need for a more balanced economics, will succeed in arousing the interest of both students and specialists in a wider economics, and educating them in that economics.

MICHIO MORISHIMA

# Foreword: Joint Production in Perspective

Lionello F. Punzo

It is fair to say that Joint Supply or Supply at joint cost belongs to the Neo-classical terminology and reflects its non- linear treatment of production functions, while Joint Production is typically associated with the Classical approach and its modern version, the linear production model. In this presentation, however, for simplicity, the two terms will often be interchanged. Until recently, the topic has been confined to article sections and book chapters, often to appendixes, dedicated to the applications and extensions of models without it. This short tract by Marco Fanno, available in English for the first time over 80 years after its original publication as a supplement to the *Giornale degli Economisti*, is one of the few instances, and the only contemporary one, of a whole monograph devoted to the topic. From the title, it is a treatise in the Neo-classical tradition. Joint Production, on the other hand, has aroused a renewed interest in the 1970s and 1980s in a quite different context, in connection with the examination of von Neumann, Sraffa and neo-Austrian models of prices, capital and growth, with a 'spillover' on the treatment of Marxian theory of labour values and prices. Fanno's treatise reminds us that the history of the subject is much longer, and is well documented as a sequel of sometime hot disputes between adherents of classical and Neo-classical schools (but also among members of the same school).

As the recent debate pointed out, the mere presence of joint production (even to a limited extent) changes some of the basic properties of prices and equilibrium as described in the Classical models. The issue becomes, then, to assess its relevance. Is this 'empirical relevance' – i.e. does it lie in the fact that joint production is the general empirical property of observed production processes? Or is it, rather, in the fact that joint production provides an analytical method? Answers to these questions reflect different conceptions, for a 'method' is not the same as a 'description'. This issue presented itself in much the same terms at the turn of this century, and Fanno gives us a refreshing perspective on it.

The history of the periodically reviving interest for joint production has something special to offer. I surmise that the series of debates never reached a generally agreed solution because too many, not purely analytical, questions are at stake. However, though themes are cyclically the same, perspectives are changing, and that of Joint Production is a case study for comparing modelling strategies and appraising theoretical motivations. One may not wish to go as far as to say that the difficulties encountered in its treatment played a significant role in the abandonment of the Classical approach to the theory of value and distribution. Nevertheless, its treatment reveals aspects of method and heuristics that 'simpler' production models leave in the shadows. This is the main motivation for undertaking the present edition of Fanno's work.

The theoretical investigation in *A Contribution* is motivated by the desire to arrive at prescriptions of industrial and fiscal policies, as they would be called by modern names: taxation, monopoly and its regulation, pricing in different market structures are some of the topics discussed at length, sometimes laboriously. It is this underlying motivation that justifies, at least partially, the relatively little attention paid to the analysis of 'existence of equilibrium solutions', an attention very limited even by the standards of mathematical economics at Fanno's times. One perceives that, in the time that has since elapsed, economics – and in particular the theory of prices – have developed into a purely theoretical exercise, where reference to immediate empirical cases for motivation has been reduced or altogether abandoned. Such an approach, though born in the modern general equilibrium economics (Arrow–Debreu-style), seems to have spread to the most distant lines of thought.

Section 1 sums up Fanno's motivations and interests in writing *A Contribution* and relates them to his research project as well as to the background of his times. Section 2 discusses Fanno's definition of 'supply at joint cost' in relation to other definitions used in contemporary literature. Section 3 elaborates on the analytical method in *A Contribution*, and suggests a relationship with some features of Hicks' approach in *Value and Capital* (1939). In the final section 4, discussion of the issues of determinateness and existence of equilibrium prices will bring us back to themes that have surfaced and captured the attention in the modern debate.

## 1 ON FANNO AND HIS TIME

At the time Fanno wrote this essay, the school of Italian economic thought, once on the frontier with such thinkers as Salimbeni, Galiani and the Illuminists, had fully come back to a high figure. To quote Schumpeter: 'The most benevolent observer could not have paid any compliments to Italian economists in the early 1870s; the most malevolent observers could not have denied that it was second to none by 1914'.[1] 1914 happens to be the year of *A Contribution to the Theory of Joint Supply*.

Schumpeter kept a close eye on his contemporary Italian economists. He was well read in the Italian economics literature, like all economists of the Austrian School, and *History of Economic Analysis* gives precious hints. In his reconstruction, the landscape of Italian thought had three successive layers, corresponding to three time periods and leading figures. The period of the forerunners (e.g. Ferrara, Graziani) is followed by the period dominated by M. Pantaleoni, and finally by the time of V. Pareto. Although working under the close protection of Loria, at the beginning of his career, Fanno feels the influence exerted on the Italian economists by Pantaleoni, directing their reflections towards topics in applied economics and fiscal policy. Fanno is in a group with such people as Antonelli and E. Barone. Only later would he get closer to the Paretian school, when it became the dominant school of thought in Italy, but he always remained a loose partner.

*A Contribution* belongs to the intermediate phase between these two in Fanno's personal evolution and in the evolution of the Italian economists' milieu. The two processes largely mirror each other, notwithstanding some peculiarities in Fanno's particular intellectual *iter*: 'even independently of Pareto, Italian economics attained a high level in a variety of lines and in all applied fields'.[2]

Variety of interests together with a certain tendency to heterodoxy[3] is the distinctive character of Fanno's scientific production. His works can be grouped under three principal headings, which correspond to different phases in his research. He begins with the theory of development, and in particular the theory of colonisation. His interests move on to the theory of money and credit, where he focuses on the Wicksellian theme of the relation between discount rate and long-run rate of profit (the natural rate) and on its influence on inflationary and deflationary processes. The conclusive moment of Fanno's meditations is represented by *Teoria delle Fluttuazioni Economiche*,[4] the one of his works which received most international attention. There, economic fluctuations are treated as persistent phenomena of endogenous origin, ignited by external shocks (an idea that has some remarkable proximity with R. Frisch's theory). It is, however, remarkable that Fanno's theory is based upon the interaction of the monetary and real mechanisms, while the theories of fluctuations available at the time (and now, one can say) were unilaterally referring to either mechanism.[5]

There is also a fourth (small) group of contributions, written in a vein closer to pure theoretical speculation, which represent periodic but temporarily sparse incursions into the theoretical realm rather than a systematically pursued research interest. *A Contribution* is one of them, and is accompanied by a coordinated title, *Contributo alla teoria economica dei beni succedanei* (1926), and two other equally short monographs.[6] Only a small share in Fanno's production is devoted to pure economics: empirically motivated research and applied work have always exerted a greater attraction on him.

Schumpeter, who was educated in the German-speaking milieu with its heated debates between the Historical and the Austrian ('Logical') schools, could not fail to appreciate the distinctive feature of the Italian economists of the turn of the century: their ability to marry economic theory and economic history, analysis and application. Fanno is an empirical or applied economist of high rank and this shows even when he does theory, as in *A Contribution*.

This provides a key to understand the motivations for and the approach taken in this work.

## 2 FANNO AND MARSHALL

A specialised monograph, *A Contribution to the Theory of Joint Supply* is an exception in the literature on production and supply. Its existence reminds us that the intricacies of the topic received attention in both Classical and Neo-classical literature. A casual perusal of the literature, at least in English, shows that by the time Fanno set down to write his work, Joint Supply, though still presented as a 'complication' of the standard cases of production, had filtered out of Treatises and found its way into textbooks. Two notable examples are the works of Nicholson and Flux, which did not entertain the ambitious, all-encompassing plan of Alfred Marshall's *Principles*.

Marshall's *Principles* provides the point of departure and theoretical backdrop of Fanno's own investigation, which is therefore fully inserted in the Neo-classical approach. One can reconstruct two motivations for his undertaking. The first is to fill in conceptual gaps in the existing analyses (the reference is explicitly to Marshall) and to provide a general treatment of joint supply; then to integrate it into a general treatment of the 'pricing problem'.[7] The latter is the ultimate theoretical construction which is aptly called 'organic' (Fanno, 1925). Realisation is planned in successive stages, by implementing a procedure of progressively adding successive 'theoretical bricks' which recalls the analytical structure in Walras' *Elements* (1954) and the techniques of classification of the biological sciences at the origins.

To Fanno, *some* of the goods are complementary in production, and joint supply is *not* the general case that the literature should have treated at greater length but was neglected. Goods produced in this 'regime' span one group in a two-parameter family classified according to the principles of complementarity and rivalry, on one side, and demand and supply, on the other.[8] Goods can be jointly supplied or jointly demanded, *and* can be complementary or substitutes to each other. Only the first are the subject of this monograph, which therefore offers only a partial treatment of the pricing problem. Nevertheless, the analysis of the general movements of prices and quantities through the interconnections between their markets is Fanno's final objective. Before attacking this, one needs to consider all the other, 'dual goods' in the matrix classification above: there will be a separate study of the correlations existing between the price behaviours of goods in each group, under the assumption of having decoupled the sets of price equations that belong to distinct groups of goods.

Goods' classification is a rhetorical device to simplify the treatment, as their qualitative spectrum varies gradually; borderlines are fuzzy. In the second phase, this 'logical' device has to be abandoned to 'invert the initial process so as to reconstruct the system as a single whole' (Fanno, 1914, p. 13). Concatenation between the two analytical phases looks simplistic as the latter is built upon the *imperfect* and *incomplete* laws derived in the former. The realisation of the overall objective requires, in the final phase, the coordination of the general theory of determination of relative prices with the theory of monetary circulation. This preoccupation, constantly at the centre of Fanno's theoretical reflection on economic dynamics, occurs here, too. The interplay between monetary and real variables, elaborated from Wicksell, links *A Contribution* to the theory of business fluctuations, which soon came to dominate the author's theoretical reflection.

Fanno realised only the first of these planned stages of research, and even this one only partly. The counterpart to *A Contribution*, dedicated to substitutable goods, found the light eventually in 1926; the overall plan was never accomplished. Fanno's interests, which one sees already veering toward dynamics, shifted to the analysis of business cycles, which must have appeared more promising to an empirically-minded economist.

The starting point of *A Contribution* is the remark that existing treatments of joint supply consider only special cases: the single isolated market, which is in a competitive regime and closed to international trade. Fanno sets out to extend the received theory to cover monopolistic regimes and internationally open markets. This variety of combinations of market structures and inter-market relations is first analysed by means of the *partial equilibrium* approach, focusing on the (closed–open, competitive–monopolistic) market for a single good. By its very nature, however, the working of the price mechanism of jointly produced goods cannot be fully understood in isolation from one another. It is therefore, planned to be further generalised to any number of inter-related markets.

There is thus, a twofold reason for the analysis in *A Contribution* to be part of a planned complex theoretical architecture: it refers to one type of price correlation *and* it resorts to Marshall's technique to derive building blocks. For Fanno, just as for Marshall, the treatment of joint supply gives one tool to construct a more general model, for 'in the small' it captures the intricacies of a construction that has to reflect the working of all markets. For both authors, the problem arises of their relationship with the general equilibrium approach of Walras and Pareto: in a sense, the problem is even more central to an understanding of Fanno's mediation between the English and the Italian schools.

One need not go all the way with Schumpeter, who tends to overstress the Marshallian inspiration in Fanno's work. The suggestion that their treatment of joint supply be conceptually complementary hides important differences, perhaps not perceived clearly by Fanno himself.[9] Most of Marshall's attention is devoted to the case of attributable costs where, due to substitutability or rivalry in production, cost can be imputed to individual goods. To put it in Schumpeter's words, for Marshall, joint supply might be the *ladder* to the General Equilibrium analysis[10] built upon the principle of substitution. It 'opens up one of the several possible roads toward the recognition of the universal interdependence of economic quantities'.[11]

Fanno, on the contrary, sees Joint Supply as one manifestation of the general phenomenon of functional interdependence, and his investigation of the dynamics of the price mechanism begins with this as a 'case'. He shows his firm belief in the correlation between all prices and universal interdependence of all markets when he anticipates that the laws regulating equilibrium displacement will result from the networking of markets. His ambitious plan is conjugating Marshall's with Walras' (and Pareto's) viewpoints, in an operation which might have appeared theoretically suspicious at the time,[12] but had a fundamentally operational motivation. Both components had a role in his synthesis, the Walrasian one provided the 'vision'.

Marshall takes up joint supply to show how his approach to the imputability of production costs could be extended to cover that case. For Fanno, borrowing the Partial Equilibrium approach is a 'practical' means to *decode* general interdependence. It is in this specific sense that the sentence 'The extent to which partial and general equilibrium analysis may cooperate shows up well in Fanno's 1914 and 1926 contributions'[13] captures Fanno's attitude in *A Contribution*. In the seemingly simplified mental experiments constructed *in vitro* in this little treatise, he saw (or, rather, perceived) the possibility of finding a path to reveal the building blocks of more general laws. This belief in a sort of linear and additive relation between partial–small and general–big, upon which a practical approach to general equilibrium could be founded, proved historically to be ill-founded, as the developments of the General Equilibrium literature soon showed.

## 3 JOINT SUPPLY

A glance at the 1913–14 issues of the *Quarterly Journal of Economics* (*QJE*) suggests that there might be a second, not less important, motivation for Fanno's theoretical quest in *A Contribution*. I am referring here to the debate on railway fares and freight tariffs which saw Pigou and Taussig as opponents then, but was really the culmination of an abundant specialised literature that blossomed at the end of the century. The *QJE* dispute finds an echo in the opening section of Fanno's work, in his discussion of the choice of an appropriate definition for *supply at joint cost*.

On the *QJE* pages, the joint supply issue was tackled in relation with a practical problem: devising an appropriate pricing scheme for *services jointly offered* (like transport of different minerals to different final purchasers). Railway transportation offered the immediate inspiration and the practical ground of observation. The assessment of the presence of monopolistic behaviours and of regulatory implications were the focus of the exchange between Taussig and Pigou.

Beyond the apparently conciliatory words of Pigou bowing to Taussig's authority their views and attitudes were profoundly different, the former looking for a clear cut 'technological' definition (one would say an axiomatic statement), the latter for an economic intuition.

For Taussig, the conditions for joint production of goods and services arise whenever there is large fixed capital in operation – i.e. plant and equipment – which is able to generate a plurality of output over the same time period or at the same date. Such form of indivisibility at the level of capital stock translates into the impossibility of reckoning the individual cost components for the various items in the output bundle. In fact, in the operation of railways

> The plant – almost all of it – is used for the traffic as a whole, without the possibility of allocating any part of this constituent part of supply price to one or another item or class of traffic... Now I am convinced that the conditions of supply for railway transport are preponderantly joint.[14] Evaluating the contribution of capital investment to the individual components in a vector of output flows has as dual problem, that of evaluating the contribution of an investment over its (economic and/or physical) lifetime to the individual components of a time series of output.

If the former, or 'horizontal' joint production is typically associated with the presence of large equipment and production facilities, then it derives from the economic decision to undertake the corresponding investment. The regime of joint cost production, in other words, is a result of optimising decisions taken after assessing the relative profitability of producing with different types (sizes) of equipment. In other words, it is not the result of mere technological indivisibility.

In Pigou's view, on the contrary, one should talk of joint cost only

> whereas it is impossible to increase the supply of cotton fibre without at the same time increasing the supply... of cotton seed, it is not possible to increase the supply of transport services for copper without at the same time increasing the supply of transport services for coal.[15]

The implications of accepting Taussig's view, for both positive as well as normative economics, are far-reaching, and they are clearly stated by his opponent. If the services of transporting passengers and goods were produced jointly, ‘even under conditions of free competition, different rates would normally be charged for them’. On the other hand, charging different tariffs for what was the same service, would represent discriminatory behaviour, that could be based only upon monopolistic power. In fact, if ‘classification’, ‘discriminating rates’, charges based on ‘value of service’, are the results essentially of monopoly, as Pigou contends, they must go with the inhibition of monopoly exploitation. But if they are the results mainly or largely of joint supply, they may remain ‘even tho the railways arc compelled to conduct their traffic on the same principles as competitive industries’ (as Taussig recognised[16]). The implications for what we would now call industrial and in particular regulatory policies (e.g. anti- monopoly legislation) are the fundamental motivation behind a debate which at first sight seems to revolve around the purely theoretical issue of choosing an appropriate definition. The exchange, moreover, reveals two different conceptions and an ambiguity which runs through the history of the subject all the way down to recent years. This is due to the simultaneous presence of a technological and an economic definition of joint supply.

Pigou tried to clear all this up by introducing the distinction between *supply at joint cost* and *supply at common* (or *shared*) *cost*. The former would reflect the technical conditions of production (i.e. properties of production functions) and broadly corresponds to the modern usage when we talk of intrinsic joint production. The latter reflects the economic convenience that leads to organising production in such a way as to obtain a vector of outputs, i.e. a form of product diversification. But Pigou’s attempt at clarifying the terms of the question is only partly successful. Though rigorous, his distinction is often blurred when one comes to apply it to practical cases.

Joint cost in the Pigouvian sense is certainly common cost, but not vice versa.[17] Having made this remark, Fanno rephrases the distinction as one between a strict (Pigou) and a broad (Taussig) definition. Taking Pigou’s side, he insists on reserving joint cost for the cases where it ‘occurs spontaneously’ while in the broader definition, it emerges ‘purely [as] a result of considerations of convenience’.[18]

Taussig, however, is not taking advantage of a definitional ambiguity when speaking of a *principle of joint cost* which ‘may be applicable even tho a supply of one does not necessarily entail the supply of the other’ good.[19] He is touching upon a real difficulty, both a theoretical and operational difficulty, which only an altogether different, axiomatic, approach could overcome. Pigou and Fanno took 'wheat and straw' as their paradigmatic example. As a shortcoming of their logical rigour, joint supply was restricted to what we call 'intrinsic joint supply', perhaps the less economically interesting case. Pigou, and Fanno with him, left out fixed capital which is the case that, with a certain amount of confusion, Taussig had in mind. He was stressing a 'horizontal dimension' of joint production associated with fixed capital, which is dual to the 'vertical dimension' at the centre of Austrian capital theory. Pointing out that both of them stem from the fact that there are economic decisions behind production with fixed capital goods, Taussig was confusedly thinking of what came to be known as the problem of Akerman (and Wicksell).

Fixed capital goods are producers' goods whose useful lifetime is longer than the unit production cycle conventionally chosen. Calculating the cost and the return from their operation poses an imputation problem. At the end of each cycle (e.g. the crop rotation year), they reappear as only 'partly worn out machines', until they are finally scrapped. Therefore, at any date in a given utilisation profile, the 'machine' enters as input at the beginning and must be reckoned as output (of one-period-older machine) at the end, together with the intermediate or final commodity (or commodities) it was employed to produce. This is one aspect of the evaluation problem, its 'accounting side', for the operating lifetimes of machines are given. In this form it was tackled by the Classics.

However, the economic lifetime of a machine is itself the result of a choice: depending on the economic environment, it may turn out to be convenient to dispose of a machine before it has become physically useless. Hence, with lifetime an endogenous variable, only producers' goods which are chosen to be operated longer than one unit period-qualify as fixed capital goods, and such unit period of reckoning becomes an endogenous variable, too. It becomes the time span needed to produce identifiable output with capital goods of the shortest life. In accounting, on the contrary, the set of fixed capital goods (and the accounting period) is taken as a datum, the result of past decisions. Thus, if we abandon the accounting viewpoint, the analysis of fixed capital is associated with that of economic decisions made in respect of technological constraints.

If we can re-read it with the modern eyes, after von Neumann's model of growth and prices, Taussig's contention was not ill-founded, after all. The operation of fixed capital entails *per se* aspects of joint production that could (rather, should) be treated on the same footing as intrinsic joint production. Could we re-phrase Taussig's argument this way, he surely would be on the better side with respect to Pigou and Fanno. As worked out by von Neumann and Sraffa, well after Fanno wrote his monograph, joint production has become an analytical method.[20] The method of joint production is the descendent of supply at shared cost, just as intrinsic joint production claims Pigou's joint cost as ancestor. However, they do not relate to each other as general to particular cases do.

The attention of the economists has been dwindling from one to the other. Edgeworth's definition in *Palgrave* (1910b) is an authoritative example of a wording that both Pigou and Taussig would have agreed with when he says that joint supply takes place 'when two or more things are produced by one and the same process, so that the expenses of producing them altogether are not greater than the expenses of producing one of them alone'. The Pigou–Taussig debate (and I dare say, Fanno's immediately subsequent – *A Contribution*[21]) did not help to resolve the opposition between *method* and statement of a *technological property*. Moreover, if the former is relevant as far as production with capital is considered, the latter may or may not be empirically relevant,[22] or it may be theoretically relevant and practically irrelevant.[23] Arguments supporting the relevance of joint production in the two conceptions are different.

I believe that neither Classics nor Neo-classics ever made up their minds on this issue. The reason for this lies in the fact that neither school of thought adopted a deductive approach, where technology is axiomatised as part of primitive data. They both start from intuitive observations, instead, and take realised technology rather than the book of blueprints. In this *ex post* view, the distinction between the two kinds of joint production becomes blurred. The mere fact that a certain technological operation is observed is sufficient for its 'necessity'. Technological necessity and economic convenience cannot be separated from one another.

This is why Taussig never yielded to Pigou's otherwise compelling arguments. Axiomatic was anyway beyond the theoretical mentality of early 1900s economics. It was next step, to be made by J. von Neumann. Not everything was resolved, however, as we still meet with Sraffa's sentence, 'Joint production is the genus of which fixed capital is the leading species',[24] where again 'method' and 'empirical description' are assimilated. At the other end of the century, we have not made much progress towards clarification.

## 4 FANNO AND HICKS

Fanno's adoption of the restricted definition advocated by Pigou is a shortcut to identifying and isolating one group of phenomena, joint production of the 'intrinsic' or 'technological' type. It does not imply the decision of axiomatising (Fanno could not be more remote from this idea), nor does it carry Jevons' implicit value judgement, that joint production is the general phenomenon in production. The study of the conditions for the establishment of equilibrium prices for joint products belongs to a different plan: providing benchmarks for assessing observable economic behaviours and evaluating policy measures. Therefore, via comparative statics, the core of the monograph states a set of propositions on the effects of disturbances on the supply (due to taxes or subsidies) and/or on demand sides. They are not simple 'corollaries':

> To know the laws of prices really well, it does not suffice to affirm that all prices are connected among themselves. It does not suffice to furnish the system of a group of equations in which the prices of all goods and services function as simultaneous determinants. In addition, it behoves [us] to consider the nature of the relationships that exist between prices – the direction and the intensity into which various prices move one with respect to the other – the modes in which the various groups of prices are connected. Concerning all this the school of Lausanne has until now kept silence.[25]

The grandeur of the vision proposed by Walras and Pareto, with 'a vast, complex, tightly-knit web of relations holding all prices together by their common chain of solidarity', marks also its *sterility*. To Fanno, the theory of General Equilibrium 'furnished only a scheme rather than a full co-ordinate system for the laws of prices.' A grand plan had been laid down for people's consideration, but its ultimate purpose remained elusive. Fanno's concern for an operational reconstruction of price theory shows how advanced was Italian thought, as it already perceived the need to go beyond general equilibrium theory as it stood at the time.

In Fanno's era, three possibilities were in principle open for further investigation: (1) the first was to deepen the mathematical understanding of the statements and theorems of that theory, which account for the working of the 'invisible hand'; (2) the second entailed investigating further the computability of equilibrium solutions, which could provide a benchmark in descriptive and/or normative economics; (3) the last was to work out the dynamic laws that could be used for 'economic prediction', forecasting in a broad sense. To E. Barone, among the Italians, the task of progressing beyond the initial statements of the grand vision meant following alternative (2) above. He demonstrated how Walrasian (and Paretian) analysis could be used for planning by showing the equivalence of the latter with computing the solutions to a system of equations. The elaboration of the theorems on the existence of equilibrium had to wait some 20 years to be tackled in earnest, in Vienna first, and finally in the USA.

Fanno tried out tracks (3), and a number of Italian mathematical economists who were active until the 1940s, went along with him, Giulio La Volpe among them. In his investigation, Fanno deploys a method that, later, would be called by Hicks the static equilibrium method. He makes the twofold assumption, that there is a known equilibrium to start from (without investigating the mathematical conditions for its existence), and that such equilibrium is described by the equations of his model. The sentence 'Consider a given equilibrium and suppose that the initial conditions of it change consecutively one by one'[26] reminds one of Walras' assumption of his model describing 'an equilibrium *ab ovo*'. Fanno, then, proceeds to investigate the working of the 'invisible hand' not in establishing consistency among individual choices, but in recreating harmony between them. A static method was used to tackle dynamic issues.

Assuming the existence of an equilibrium without proof is obviously insufficient and no longer appears legitimate to our eyes. The assumption, however, is purely instrumental, to focus on the adjustment processes and on the laws of equilibrium displacement, and it may generate a knowledge basis for prediction and normative prescription. This is the result that Fanno was striving to obtain; assumption seemed to be the price for pay for it. The shortcoming of such procedure is that the resulting analysis has to remain 'local' and may consider only shifts in parameters and environment conditions which do not lead 'too far away' from the original equilibrium. This, however, is what Fanno's analysis intends to do.

Even before the Arrow–Debreu–McKenzie codification, General Equilibrium theorists were aware of these limitations, and accepted them for a purpose, rather than simply for their mathematical inability to handle the general problem.[27] Global analysis of system properties was traded off for a more desirable result, obtaining dynamic laws that in principle could be explicitly derived, though perhaps only qualitatively. At the time, to build a general analysis only meant relying upon the principle of universal functional interdependence. Fanno tried to marry Marshall with Walras and Pareto; some time later, Hicks attempted the compromise with Keynes' thought which prepared the Neo-classical Synthesis. Fanno's inspiration is similar to the one behind Hicks' project in *Value and Capital*.

Hicks' defence of his own modelling strategy was that models were mental laboratories constructed to carry out conceptual experiments. Reading Fanno's *A Contribution* conveys very much the same impression. Fanno, too, is building an artificial laboratory for experiments that our discipline could not otherwise carry out at the time. New technologies available only recently at reasonable cost and on a mass basis seem now to promise different, life-size laboratories for our experiments of the future.

The reader may or may not believe in the theoretical and/or empirical relevance of Joint Production. However, the debate about Joint Production at the beginning of this century offers an unusual perspective on the unfolding of a research project, that originated in an attempt of exiting from the theoretical confines of general equilibrium economics. In a sharp departure from this, much of the recent debate has revolved around those issues of equilibrium existence and determinateness which are typical concerns of the modern general equilibrium theory.

## 5 EQUILIBRIUM EXISTENCE AND DETERMINACY

Though Fanno is careful in dealing with two interdependent markets for jointly produced commodities, he boldly bypasses the mathematical issues arising in the general case with *n* markets. He seems to believe that conditions for equilibrium existence for two commodities carried over to *n* commodities without relevant modifications. The recent debate has shown that this is not true and has focused on the role of 'demand' when the system of price equations is not fully determinate. One tends to re-read old treatments of joint production with modern eyes, generating misunderstandings that need to be spelled out.

In the standard Neo-classical approach, the number of production processes available *ex ante* is always infinite due to the possibility of continuous substitution between inputs along smooth production functions *and* variable returns to scale. With each producible level of output is associated a process (possibly, a set of processes) with a unit cost or supply price. Therefore, technical conditions of supply alone are not able to determine prices for any commodity: they yield 'too many' price equations from which to choose.

On the other hand, the number of producible goods is finite, and hence smaller than that of technological processes. Thus, as the determination of equilibrium prices cannot be separated from the determination of output levels, 'demand conditions' are introduced to pin down which process(es) will be activated (at the chosen levels of output), *and* the corresponding vector(s) of equilibrium prices. 'Solving' the market equilibrium equation between demand and supply implies a notional search procedure through an infinite set of supply and an infinite set of demand values, to find pairs of demand and supply relations that hold as equalities. In equilibrium, equations holding true are (almost) never fewer than the prices to be determined.

Therefore, contrary to a frequently entertained opinion, demand is not brought into the Neo-classical theory of production to supply some 'missing equations'. Its role is to choose a subset of relations from a system which, anyway, remains in general mathematically over-determined.[28] Such over-determinacy implies that the equilibrium technique is not uniquely identified: several techniques are normally eligible for the production of the same set of goods. Qualifications to this general rule must be introduced if we assume the special case of linear fixed coefficient production functions. The principal one is that the choice of the set of processes to produce any one good in equilibrium becomes independent from both the scale of production and the composition of final demand. Equilibrium prices become independent of final demand composition. This is nothing but the Non-Substitution Theorem, a particular case of the General Equilibrium pricing model.

In Neo-classical analysis, therefore, the twofold crucial role played by demand is due to the generally assumed non-linearity of production functions. Hence, demand was not introduced to meet the difficulties encountered by the Classics in the treatment of joint production (as suggested by Kurz, 1986): demand was already there, before and independently of joint supply. Its presence 'may help', but it does not solve the problem: how to allocate the aggregate cost of two or more jointly produced goods which have distinct markets. The Neo-classical approach saw it as a problem of imputation of value, in other words as a problem of 'disaggregation'. This motivates the discussion of the derived demand function in Marshall (Book V), and it is even more clear in Fanno's treatment which is based upon an ingenious assumption of fixed proportions between component outputs.[29] Hence, no opposition in principle between production with and without joint production was felt, as pointed out by Edgeworth:

> The value of joint products is regarded by Mill as a 'peculiar case' apparently because it varies with the demand for the products;... But with the wider conception of cost of production varying with the quantity produced (Sidgwick, *Pol. Econ.*, Bk ii, ch. ii, s.7; Marshall, *Principles*, *passim*) this difference between single and joint products disappears.[30]

The complementary role assigned to demand, on the other hand, demonstrates in a static framework the impossibility of decoupling two parallel mechanisms of dynamic adjustment. Decoupling is possible when the conditions for the Non-Substitution Theorem hold. But such conditions are very restrictive, while in the General Equilibrium approach it is natural that quantities and prices be determined simultaneously, as interdependent variables. Hence, it is at its best with non- linear demand and supply functions, with variable returns to scale considered the general case and constant returns a particular realisation. Joint production, even on a limited scale, shows that the latter are a necessary but not sufficient condition to decouple the price from the quantity equations.

The Classical approach and its modern neo-Ricardian re-formulation rely on a constant coefficient technology and encounter the possibility of finding fewer equations than price unknowns. In fact, with a discrete set of processes, a process with multiple outputs introduces degrees of freedom which need be matched by some other equations to obtain a determined system. The criticism levied against the Neo-classical approach can, therefore, be laid at the Classical doorstep: the Classical system must be 'closed' if *Joint Production* is introduced; the Neo-classical system (with production) must be closed always – i.e. independently from the specific properties of the output vectors. In the Classical approach, however, the reason for this is the lack of data.

In fact, that the technology matrix lists more (or, at least as many) processes than goods is not sufficient to guarantee that there be at least one square submatrix for a 'technique adopted in equilibrium'.[31] If it is square, the Classical price problem with joint production is solvable in the same way as with single-good production, prices being uniquely determined by production conditions alone.[32] (This shows that the Non-Substitution Theorem may be generalised in certain cases.[33])

Put in this way, the issue becomes: how likely is it that the equilibrium submatrix turns out to be square? In this form, it has been discussed in the recent neo-Ricardian literature.[34]

The problem, however, is wrongly located at the level of the properties of the solutions, while it arises with the data or technology matrix. If the technology matrix has $m$, the number of rows/goods, *greater* than $n$, the number of columns/processes, the solution matrix can never be square. In this case,[35] either more processes are 'added' to the data set; or demand has to be brought in; or, finally, some goods must be 'dropped'. 'Adding processes' is the solution attributed to the Classics and adopted also by Sraffa. In a certain sense, adding demand conditions (though in a peculiar form) is von Neumann's own solution. However, increasing the number of columns to construct a square data matrix might be as good as decreasing the number of rows. It can be argued that, if the number of goods exceeds the number of processes, it is mathematically unlikely that all goods receive a positive valuation in any equilibrium solution.[36]

A group of 'candidate-free goods' is therefore any subset of $(m - n)$ goods selected out of $m$ producible goods. As there is a finite number of such groups, the treatment of the equilibrium pricing problem may proceed by searching through the set of square submatrices obtained by deleting the corresponding rows in the data matrix. Barring degeneracy, at least one of these matrices will be associated with equilibrium. This is the solution implemented in the treatment of ageing fixed capital goods – i.e. when joint production is a method. Assume that this is the only type of joint production and, moreover, that once installed machines (or better, 'plants') are not transferable. Then, for each 'machine' and a fixed lifetime a square submatrix is obtained with processes that employ it, the last year of life signalled by its being treated as a candidate-free good (and getting a zero coefficient). All such submatrices are obtained by dropping selected old machines from the list of commodities and they are square; so the equilibrium matrix will be square, too. We have to solve a set of determinate systems of price equations.

In a certain sense, this alternative 'solution' seems to be close to the Neo-classical treatment of Walras, Pareto and Hicks, that Fanno also follows in his *A Contribution*. It is a rationalisation of the fact that these authors assume an equilibrium and study some of its properties. If we are in equilibrium, the list of free goods is known.

This approach has both advantages and disadvantages. On the latter count, it hides the optimisation process in the choice of techniques, which instead emerges in an inequality approach with free goods endogenously determined. On the former count, the linear treatment of joint production is traced back to the spectral analysis of linear operators, which is the fundamental algebraic theory of matrices. Through the latter, pricing for joint production links up with the linear theory of dynamical systems.

# Foreword: Joint Production and Joint Consumption

Michio Morishima

## 1

The problem of joint production which Marco Fanno is concerned with in this volume may be discussed from a number of points of view. First we may deal with the problem of determination of prices of joint produces in the framework of general equilibrium that works under the condition of perfect competition. In a particular case in which these joint produces are also jointly consumed, the excess demand functions of these commodities are closely related to each other through the coefficients of joint production and joint consumption; they form a group of intrinsic complementary consumer goods. Thus we see that the problem of joint production is connected with the problem of complementarity.

Secondly, we conceive of two economies, each containing two groups of commodities. In economy $A$, commodities in either group are substitutive for each other but complementarity prevails between the groups. On the other hand, economy $B$ has two competitive, and hence substitutive groups, within each of which complementarity prevails. In $A$, if commodity $i$ is a substitute for $j$ and $j$ is also a substitute for $k$, then $i$ is a direct substitute for $k$, because $i$, $j$ and $k$ all belong to the same group, while if $i$ and $k$ are both complementary with $j$, $i$ is a direct substitute for $k$, because both $i$ and $k$ belong to the group to which $j$ does not and, therefore, the former two belong to the same group. Similarly, in $A$ if $i$ is complementary with (resp. substitute for) $j$, while the latter is substitutive for (resp. complementary with) $k$, then $i$ is directly complementary with $k$. Thus $A$ satisfies the conditions: substitutes of substitutes or complements of complements are direct substitutes, and substitutes of complements or complements of substitutes are direct complements.

On the other hand, we may confirm that the following rules hold in $B$: substitutes of substitutes or complements of complements are direct complements, and substitutes of complements or complements of substitutes are direct substitutes. These rules are entirely opposite of those for *A* above, so that *A* and *B* may be referred to as an *M* and an anti-*M* system, respectively, and may be considered to form conjugate regimes.

It has been shown that the *M* system is an extention of the well known gross-substitute system and obeys the same laws of repercussion of prices. But the anti-*M* system has so far been left unexamined, though there are some conjectures made by Hicks (1939, pp. 74–5, 105–6). I take up it below in such a particular case that the system has two joint production groups; especially it is shown that certain clear-cut laws follow in the extreme case that these two groups of commodities are not only jointly produced but also jointly consumed groupwise. Of course, the laws thus obtained are not general enough, but may be regarded as a prototype for more satisfactory investigations in the future.

Finally, Fanno examines joint production under monopoly, too. There is, however, no significant difference in the results. It is true that the level of joint production and its 'global' price are different under monopoly from that under free competition. Nonetheless, the main results obtained below – that is, the conclusion that the distribution of the global output price among the prices of the elements is uncertain and some of them are offered gratuitously or discarded by making them free goods – do remain intact. It is shown that these are true even though the number of rival producers increases. This means that these laws of prices are independent of the structure of the market but characteristic of joint production and joint consumption.

## 2

Let us begin with the first point. The work by Fanno on joint Production may be regarded as a precedent for the comparative statics analysis of repercussion of prices based on the general equilibrium theory. This area of research is more interesting to us now, because Hicks, the leader in this field, did not actually examine effects of a change in the condition of supply of a commodity upon the prices of its joint outputs, whereas Hicks' production function is formulated in such a general form that it can deal with joint supply. In fact, in Hicks (1939) he could definitely comment on the effects of an increase in the demand for one commodity upon prices of other commodities only in the case of the above-mentioned rules of a chain concerning substitutes and complements for the $A$ system being fulfilled.[1] Under these two sets of rules, the system may be reduced to the one of gross substitutes – that is, the one in which all commodities are gross substitutes for each other, no complementarity being present.

This is seen in the following way. Where the chain rules hold, it can be shown (see Morishima, 1952, pp. 101–13) that the matrix of partial derivatives $(\partial E_i/\partial p_j)$ of excess demand functions $(E_i)$ with respect to prices $(p_j)$ may be arranged to the pattern of signs:

$$\begin{bmatrix} - & + & \cdots & + & - & - & \cdots & - \\ + & - & \cdots & + & - & - & \cdots & - \\ \cdot & \cdot & \cdots & \cdot & \cdot & \cdot & \cdots & \cdot \\ \cdot & \cdot & \cdots & \cdot & \cdot & \cdot & \cdots & \cdot \\ \cdot & \cdot & \cdots & \cdot & \cdot & \cdot & \cdots & \cdot \\ + & + & \cdots & - & - & - & \cdots & - \\ - & - & \cdots & - & - & + & \cdots & + \\ - & - & \cdots & - & + & - & \cdots & + \\ \cdot & \cdot & \cdots & \cdot & \cdot & \cdot & \cdots & \cdot \\ \cdot & \cdot & \cdots & \cdot & \cdot & \cdot & \cdots & \cdot \\ \cdot & \cdot & \cdots & \cdot & \cdot & \cdot & \cdots & \cdot \\ - & - & \cdots & - & + & + & \cdots & - \end{bmatrix} \tag{1}$$

where $\partial E_i/\partial p_j$ takes a positive sign or a negative sign according as good $i$ is a gross substitute for or a gross complement with good $j$; in the case of $i = j$, it takes on a negative sign under a usual set of assumptions. It is noted that in (1) the column and row for the numeraire commodity $n + 1$, $\partial E_i/\partial p_{n+1}$ and $\partial E_{n+1}/\partial p_j, i,j = 1, \ldots, n$ are omitted. (They are usually positive.) The matrix (1) shows that the commodities, $i = 1, \ldots, n$ are grouped into two sets such that they are gross substitutes within either of the two, and gross complements between them.

Let the number of commodities belonging to the first group be $m - 1$. Let it be $v$. Multiply the matrix (1) by $(-1)^{2v}$. Then it may be written as

$$\begin{bmatrix} - & + & \cdots & + & + & + & \cdots & + \\ + & + & \cdots & + & + & + & \cdots & + \\ \cdot & \cdot & \cdots & \cdot & \cdot & \cdot & \cdots & \cdot \\ \cdot & \cdot & \cdots & \cdot & \cdot & \cdot & \cdots & \cdot \\ \cdot & \cdot & \cdots & \cdot & \cdot & \cdot & \cdots & \cdot \\ + & + & \cdots & - & + & + & \cdots & + \\ + & + & \cdots & + & - & + & \cdots & + \\ + & + & \cdots & + & + & - & \cdots & + \\ \cdot & \cdot & \cdots & \cdot & \cdot & \cdot & \cdots & \cdot \\ \cdot & \cdot & \cdots & \cdot & \cdot & \cdot & \cdots & \cdot \\ + & \cdot & \cdots & + & + & + & \cdots & - \end{bmatrix} \qquad (1')$$

This shows that Hicks' economy satisfying the chain rules for the $A$ system is convertible into a system of gross substitutes, so that in this case the presence of complementarity does not lead us to a problem that needs a new mathematical theorem to solve it. The system is essentially the same as the one of no complementarity. Obviously, it is easily confirmed that an economy consisting of a group of several kinds of tea and another one of various sorts of sweetening materials satisfies the chain rules. As we see, Ceylon tea that is a substitute for Assam tea is a substitute for Earl Gray tea which is a substitute for Assam. Also we see that sugar, a substitute for honey, is a substitute for saccharine that is another substitute for sugar. Thirdly we see that sugar, a complement of Ceylon tea, is a complement of Assam that is a substitute for Ceylon. Finally, we see in the same way that sugar as a complement of Ceylon tea is a substitute for honey that is another complement of Ceylon tea.

With the sign pattern (1) of the partial derivative matrix of the excess demand functions, Morishima (1952) has shown that an increased demand for commodity $i$ raises the prices of those commodities $j$ that are substitute for $i$ and lowers the prices of those $k$ that are complementary with $i$. Though these results may be referred to as '$M$-properties', they are no more than a corollary of Hicks' theorem concerning the gross-substitute system. This is immediately clear because a congruent transpose of (1) may be put in the gross-substitute form (1').

However, this is no more than an example of complementary commodities appearing in the competitive market. It cannot be said to be even the most probable or most important case. We may at once think of another example that may be taken as being at least equally important or frequent. Consider a number of commodities which are useful only if they are consumed in certain fixed proportions. If they are, say, two sets of such commodities, complementarity prevails within each set, but the two sets may be substitute for each other. In this case, the matrix $(\partial E_i/\partial p_j)$ is of the following sign pattern:

$$\begin{bmatrix} - & \cdots & - & + & \cdots & + \\ \cdot & \cdots & \cdot & \cdot & \cdots & \cdot \\ \cdot & \cdots & \cdot & \cdot & \cdots & \cdot \\ \cdot & \cdots & \cdot & \cdot & \cdots & \cdot \\ - & \cdots & - & + & \cdots & + \\ + & \cdots & + & - & \cdots & - \\ \cdot & \cdots & \cdot & \cdot & \cdots & \cdot \\ \cdot & \cdots & \cdot & \cdot & \cdots & \cdot \\ \cdot & \cdots & \cdot & \cdot & \cdots & \cdot \\ + & \cdots & + & - & \cdots & - \end{bmatrix}$$

which cannot be transformed into the form (1′), so that Hicks' theorem is not applicable to this case with an 'anti-*M*' matrix (see Morishima *et al.*, 1973, pp. 66–7)).[2] Obviously Hicks' theorem is of no use for anti-*M* systems and we have so far obtained no definitive results for them.

My work on complementary goods has been deadlocked since 1973, because I could not develop any useful piece of mathematics that is applicable to the anti-*M* system or any other one containing complementary goods. Fanno has, however, revived my interest in this longstanding, unsolved knotty problem. I do not say that it has been solved now; my tentative approach presented below is still far from satisfactory. Although it seems to include no logical fraud, it is at least not very penetrating mathematically. Nevertheless, it is true that the work carried out by Fanno enables me to fill some of lacunae which I have found in the comparative statics analysis concerning price repercussions of the anti-*M* type.

Fanna's joint supply theory is applicable to the analysis of production of meat, roe, skin and bony parts of salmon; they are of course produced in fixed proportions. Various parts of cod are also similarly supplied in fixed proportions. Where the price of a whole cod (which is called its 'global' price by Fanno) increases, then the fishing activity is shifted from salmon to cod, so that they are substitutive on the supply side: meat, roe, skin and other parts of cod are produced proportionately. Assuming that the demand for salmon and that for cod are both inflexible with respect to changes in their prices, the matrix of partial derivatives of the functions of excess demands for salmon and cod may be written in the form of (2), in which the block for salmon (or cod) consists of four components – meat, roe, skin, and bony parts. Thus (2) is an $8 \times 8$ matrix. We then examine effects of an autonomous change on the demand side, or on the supply side, upon prices of the commodities which form an anti-*M* system.

3

Let
$$f(y_1, \ldots, y_{m-1}, z_m, \ldots, z_n) = 0$$

be the production constraint of a single firm. *y*s are products and *z*s are inputs, so that joint production is allowed for. Let $p_i$ be the price of $y_i$ and $q_i$ that of $z_i$. We assume that proportions of $y_i$s are technically determined, so that we have $y_i = k_i y_1$ for $i = 2, \ldots, m-1$. Maximising $p_i y_i - q_j z_j$ subject to $f(\ldots) = 0$, we obtain, by the familiar procedure

$$\begin{aligned} y_1 &= y_1(P_1, q_m, \ldots, q_n), \\ y_i &= k_i y_1(P_1, q_m, \ldots, q_n), \quad i = 2, \ldots, m-1 \\ z_j &= z_j(P_1, q_m, \ldots, q_n), \quad j = m, \ldots, n \end{aligned}$$

where

$$P_1 = p_1 + k_2 p_2 + \ldots + k_{m-1} p_{m-1} \tag{3}$$

that is the global price of the joint outputs. $C = \Sigma q_j z_j$ is the joint cost of production that is to be shared between outputs, $p_1 y_1, \ldots, p_{m-1} y_{m-1}$.

How should $C$ be distributed between various outputs? This is Fanno's old subject and is still discussed by contemporary economists. Moulin's (1959, pp. 307–11) article examines the same problem by using the game-theoretic approaches developed by Shapley, Shubik, Auman and others. This treatment, however, ignores how the global price $P_1$ of the composite commodity has to be distributed between prices $p_i$ of the elements of the composition, as (3) shows. How do prices $p_1, \ldots, p_{m-1}$ of joint output $y_1, \ldots, y_{m-1}$ respond to the demand for them? Unless this problem solved beforehand, it will be impossible to deal with the problem of sharing joint costs.

Suppose now the total demand for commodity $i$ is denoted by $X_i, i = 1, \ldots, m-1$, and that for $j$ by $Z_j, j = m, \ldots, n$. Evidently $Z_j$ is the sum of $z_j$s over all firms. We write the total supply of $i$ – that is, the sum of $y_i$s over all firms – as $Y_i, i = 1, \ldots, m-1$, and the supply of other commodities $j, j = m, \ldots, n$, as $Q_j$. Then we obtain the market equilibrium conditions:

$$X_1 = Y_1(P_1, q_m, \ldots, q_n), \tag{4.1}$$
$$X_i = k_i y_1(P_1, q_m, \ldots, q_n), \quad i = 2, \ldots, m-1 \tag{4.2}$$
$$Q_j = Z_j(P_1, q_m, \ldots, q_n), \quad j = m, \ldots, n \tag{4.3}$$

Let $P_1, q_m, \ldots, q_n$ be the equilibrium set of prices satisfying (4.1) – (4.3). Let us fix $X_j, i+1, \ldots, m$-1 and $Q_i$, $i = m, \ldots, n$ at the amounts that are decided by demanders and suppliers, then these equilibrium prices are determined in the market.

Next, suppose the demand for commodity 1 is increased from $X_i$ to $X_i + a$, with no influence upon the demand for and supply of other commodities $X_i$, $i = 2, \ldots, m-1$, and $Q_i, i = m, \ldots, n$. Then it is clear that we have

$$X_1 + a = Y_1(P_1, q_m, \ldots, q_n), \tag{5.1}$$
$$X_i < k_i\, y_1(P_1, q_m, \ldots, q_n), \qquad i = 2, \ldots, m-1 \tag{5.2}$$
$$Q_j = Z_j(P_1, q_m, \ldots, q_n), \qquad j = m, \ldots, n \tag{5.3}$$

Then comparing (5.1) and (5.3) with (4.1) and (4.3), we find that $P_1$, $q_m, \ldots, q_n$ are affected, but (5.2) shows that commodities, $i = 2, \ldots, m-1$, are now free goods, so that their prices, $p_2$, $\ldots$, $p_{m-1}$, take on the value 0. This means $P_1 = p_1$, that is to say, although $m-1$ commodities, $1, \ldots, m-1$, are all produced simultaneously, the joint production costs are not shared among them but are charged exclusively to the account of production of commodity 1. On the other hand, where $X_1$ decreases by the same amount as the exogenous increment $a$, (4.1)–(4.3) continue to hold, so that there is no change in prices, $P_1, q_m, \ldots, q_n$. The global price of the composite commodity 1 is thus obtained: any set of prices of the elements, $P_1, \ldots, p_{m-1}$, of the joint outputs which satisfies (3) with value of $P_1$ is the equilibrium set of prices. Thus the prices of its elements $1, 2, \ldots, m-1$, are uncertain. Since joint cost are distributed among them in proportion to these prices, $p_1, \ldots, p_{m-1}$, the cost sharing is also uncertain.

Let us next replace constant $X_i$s by more realistic demand functions. We assume that commodities $1, \ldots, m-1$ are mildly gross-substitutive between each other as meat, roe, shin and bony parts of salmon are. Moreover, a similar assumption may be imposed upon them with respect to remaining commodities, that is to say, $X_i$s, $i = 1, \ldots, m-1$, are gross-substitutive for commodities $m, \ldots, n$. Additionally we may also assume that these commodities are gross-substitutive not only for commodities $1, \ldots, m-1$, but also between themselves. Furthermore, we relax the assumption that $Q_i$s are kept constant into the one that they are supplied, like $Y_1, \ldots, Y_{m-1}$, in fixed proportions. Under all these assumptions we may further suppose that the gross substitutability on the demand side is weak, so that it is dominated by the gross-complementarity prevailing on the supply side, within either of the group of commodities $1, \ldots, m-1$ and that of commodities $m, \ldots, n$. Note in the above that $Y_i = k_i y_1$.

Thus the matrix of partial derivatives of the excess demand functions $E_i = X_i - Y_i, i = 1, \ldots, m-1$ and $E_i = Z_i - Q_i, i = m, \ldots, n$, may be of the same sign pattern as (2). It is obviously of the anti-$M$ type. Thus, the case of joint production dealt with by Fanno provides an example of the anti-$M$ class of economy.

## 4

Where the gross substitutability is weak enough, the following comparative statics laws will be obtained for the anti-$M$ model above. They are intuitive generalisations of the rules of price changes that we have acquired in the above for the case of $X_i$s and $Q_i$s being constant. Where no substitution mechanism works between commodities, an increase in the demand for commodity 1 gives rise to an increase in its price and makes other components $2, \ldots, m-1$ free. However, where substitution takes place between commodity 1 and commodities $2, \ldots, m-1$, the prices of the latter may be kept positive, but there is no case of their prices rising more than the price $p_1$. Perhaps all or most of them will diminish, whilst the aggregate – the global price (3) – certainly increases. On the other hand, by virtue of the substitution relationships between the first composite commodity $(1, \ldots, m-1)$ and the factors of production for it, $m, \ldots, n$, there is an increase in the demand for the latter. Then the global price $P_{II}$ of the second composite commodity $II, (m, \ldots, n)$ as well as the price $p_m$ of its first component, will increase, while prices of other component, $p_{m+1}, \ldots, p_n$ may fall even to the bottom level of 0, as they may become a free good.

Since commodities 1 and $m$ are gross substitutes and commodities $2, \ldots, m-1$ and those of $m+1, \ldots, n$ are gross complements with 1 and $m$, respectively, we may alternatively paraphrase the rules obtained above in the following way:

(a) An increase in the demand for commodity 1 gives rise to an increase in its own price $p_1$ and the price of its substitute, commodity $m$ that is $p_m$.

(b) It will perhaps diminish prices $p_2, \ldots, p_{m-1}$, and $p_{m+1}, \ldots, p_n$ of the other commodities that are complementary with commodity 1 or $m$. This means that substitutes are covariant in the price movement, whilst prices of complements of a commodity $i$ change in the opposite direction (contravariant) into which the price of $i$ will do. The results for the anti-$M$ system expressed in this way are very similar to those for the $M$ system that have been already stated above. While the $M$ matrix (1) is not reducible to the anti-$M$ matrix (3), and vice versa, it is surprising to obtain the same results concerning repercussions of prices from either of them – that is, when the price of a commodity rises due to an increase in its demand, then there are covariant and contravariant movements of prices of its substitutes and of its complements, respectively.

5

Let us now turn to the consumer side. Suppose a number of commodities indexed, say, as $1, 2, \ldots, m-1$ are always used in certain fixed proportions, $k_1, k_2, \ldots, k_{m-1}$; then we say that they are jointly consumed. We define the quantity of the composite commodity as:

$$x = \min\left(\frac{x_1}{k_1}, \frac{x_2}{k_2}, \ldots, \frac{x_{m-1}}{k_{m-1}}\right) \tag{6}$$

where $x_i$ is the quantity of commodity $i$ contained in the composition. Throughout the following, we take $k_1 = 1$ for the sake of simplicity. The utility function of the consumer may be written as:

$$u(x, x_m, \ldots, x_n, x_{n+1})$$

which is maximised subject to her budget equation,

$$p_1 x_1 + \ldots + p_{m-1} x_{m-1} + p_m x_m + \ldots + p_{n+1} x_{n+1} = M \tag{7}$$

Taking commodity $n+1$ as the numeraire so that $p_{n+1} = 1$, and define $P_1$ as

$$P_1 = p_1 k_1 + \ldots + p_{m-1} k_{m-1}$$

which gives us the global price of the composite commodity. Then the equilibrium conditions are written as:

$$\partial u/\partial x = \lambda P_1,$$
$$\partial u/\partial x_i = \lambda p_i, \quad i = m, \ldots, n+1$$

Solving these we obtain the demand functions of commodities:

$$x = x(P_1, p_m, \ldots, p_n)$$
$$x_i = x_i(P_1, p_m, \ldots, p_n), \quad i = m, \ldots, n$$

In view of (6) it is clear that $x_i, i = 1, \ldots, m-1$, should be determined as:

$$x_i = k_i x = k_i x(P_1, p_m, \ldots, p_n), \quad i = 1, \ldots, m-1$$

These demand functions $x_i$, $i = 1, \ldots, n$ may be treated in the exactly same way as we dealt with $y_i(P_1, p_m, \ldots, p_n), i = 1, \ldots, n$ in the previous sections.

We can show, first of all, that provided that the supply is fixed for each commodity $i = 1, \ldots, n$, elements of the matrix of partial derivatives of excess demand functions are all negative if they belong to its submatrix consisting of first $m-1$ columns and first $m-1$ rows. Then, with the supply of the commodities all being constant, an exogenous increase in the demand for commodity 1 yields an increase in its price, which induces a proportionate decrease in the endogenous demand for any of the first $m-1$ commodities. Thus $p_1$ is adjusted such that the endogenous demand for commodity 1 is equal to that part of its supply that remains after subtracting the part devoted to the increased exogenous demand. Because the demand for remaining $m-2$ commodities, $2, \ldots, m-1$, diminishes proportionately, an excess supply appears for each of them. As we assume that the supply is not adjusted, the prices of these commodities decline to the bottom; that is to say, they now become free goods. This implies that $P_1 = p_1$.

Let us now conceive of an economy in which two composite commodities *I* and *II* are traded. The first of these comprises elementary commodities $1, \ldots, m-1$ and the second those of $m, \ldots, n$. We keep commodity $n+1$ as the numeraire. We denote the quantity of *I* and *II* by $x_I$ and $x_{II}$; they are defined as before as:

$$x_I = \min\left(\frac{x_1}{k_1}, \frac{x_2}{k_2}, \ldots \frac{x_{m-1}}{k_{m-1}}\right)$$

and

$$x_{II} = \min\left(\frac{x_m}{k_m}, \frac{x_{m+1}}{k_{m+1}}, \ldots, \frac{x_n}{k_n}\right)$$

The consumer's utility function can be written as

$$u = u(x_I, x_{II}, x_{n+1}) \tag{8}$$

By the familiar procedure for maximising (8) subject to the budget equation, we obtain

$$x_i = k_i x_I(P_I, P_{II}), \quad i = 1, \ldots, m-1$$
$$x_j = k_j x_{II}(P_I, P_{II}), \quad j = m, \ldots, n$$

where

$$P_I = p_1 k_1 + \ldots + p_{m-1} k_{m-1}$$

and

$$P_{II} = p_m k_m + \ldots + p_n k_n$$

These are of course the global prices of *I* and *II*.

Let us now make an additional assumption to the effect that the two composite commodities, *I* and *II*, are mildly gross-substitutive for each other; we then find that the $n \times n$ matrix of partial derivatives $x_i, i = 1, \ldots, n$, with respect to prices $p_j$, $j = 1, \ldots, n$, is of the anti-*M* type, like (2). This assumes constant supplies of the *n* commodities, but the economy still remains to be of the same type, even though a sufficiently mild gross substitutability prevails on the supply side. We then get the following rules concerning price changes.

Before laying down the rule, we must remind ourselves of the fact that whereas elementary commodities $i$, $i = 1, \ldots, n$, are demanded in compositions, $(k_1, \ldots, k_{m-1})$ and $(k_m, \ldots, k_n)$, each of the elements is supplied independently from other components. Then we may assume that the supply of commodity 1 diminishes for some exogenous reason. Then its price $p_1$ will rise. As for commodities $2, \ldots, m-1$, because the system is assumed to be an anti *M* one, the complementarity on the demand side dominates the substitutability on the supply side; there will then be excess supplies of them, so that prices $p_2, \ldots, p_{m-1}$ will decline.

Let us next examine how the change in the prices of *I* will affect the prices of *II*. As we assume that commodities $1, \ldots, n$ are gross-substitutive on the supply side, an increase in $p_1$ induced by an exogenous decrease in the supply of commodity 1 gives rise to a shift of production from other firms towards the firms producing commodity 1. Let *m* be the commodity, the supply of which is most affected (that is, most decreased) among commodities $m, \ldots, n$. Then $p_m$ will certainly increase. However, a decrease in the demand for commodity *m* created in this way will result in a proportional decrease in the demand for the remaining commodities $m+1,\ldots,n$ in the set $II$. Then their prices will decline. This channel of repercussion confirms that the prices of substitutes change in the same direction, while those of complements change in the opposite direction.

Of course, there are other channels, too. It follows from the assumption of gross substitutability between $x_I$ and $x_{II}$ that an increase in $p_1$ brought about by a decrease in the supply of commodity 1 will create a proportional increase in the demand for $m,\ldots,n$. Therefore, their prices tend to increase. In view of the assumption that the gross substitutability on the demand side is mild, we may consider that these increases in $p_i$, $i=m+1,\ldots,n$, are not enough to compensate the original decrease in them that we have observed above. As for $p_m$ that has originally increased, it is evident that this further increase due to the substitution on the demand side reinforces our rule of price fluctuations.

## 6

Suppose commodities $1,\ldots,m-1$ form a composite commotidy on both production and consumption sides. Let $(k_1,k_2,\ldots,k_{m-1})$ be the composition on the production (supply) side, and $(k'_1,k'_2,\ldots,k'_{m-1})$ on the consumption (demand) side. It is clear that in the state of general equilibrium the demand for each commodity cannot exceed its supply. There is a possibility that the former is actually less than the latter. When this happens, the relevant commodity is, in fact, a free good. Then we have the conditions:

$$X_i(p_1,\ldots,p_n) \leq Y_i(p_1,\ldots,p_n), \quad i=1,\ldots,n \tag{9}$$

In the following we assume prices $p_m,\ldots,p_n$ are determined in the market of commodities $m,\ldots,n$; we are then concerned only with the formation of prices $p_1,\ldots,p_{m-1}$, in the subeconomy consisting of commodities $1,\ldots,m-1$. Evidently $p_i=0$ ($i$ is a free good) if (8) holds with strict inequality ‘ $<$ ’ for $i$. Defining

$$h_i = k_i/k'_i, \quad i=1,\ldots m-1$$

we assume $h_i$ to be the smallest among $h_1,\ldots h_{m-1}$. Let us suppose that there are $h_i$s which are greater than $h_1$. Let $h_2$ be one of such $h_i$s. As

$$Y_2 = \frac{k_2}{k_1}Y_1 \quad \text{and} \quad X_2 = \frac{k'_2}{k'_1}X_1$$

we have

$$Y_2 > X_2$$

By taking $Y_1 \geqslant X_1$ (equilibrium condition) and $h_1 < h_2$ (assumption) into account, commodity 2 is a free good, that is, $p_2 = 0$. Hence only those commodities $i$ with $h_i = h_1$ are non-free.

We have already seen that $Y_i$ and $X_i$ may be written as

$$Y_i - Y_i(P_1, p_m, \ldots, p_n)$$
$$X_i = X_i(P'_1, p_m, \ldots, p_n)$$

where

$$P_1 = p_1 k_1 + p_2 k_2 + \ldots + p_{m-1} k_{m-1}$$
$$= p_1 h_1 k'_1 + p_2 h_2 k'_2 + \ldots + p_{m-1} h_{m-1} k'_{m-1}$$
$$P'_1 = p_1 k'_1 + p_2 k'_2 + \ldots + p_{m-1} k'_{m-1}$$

For those $i$ with $h_i > h_1$ we have seen $p_i = 0$, while for others $h_i = h_1$. Hence we obtain $P_1 h_1 = P'_1$. Because we may change the unit of the composite commodity on the production side such that $h_1 = 1$, we may regard $P_1 = P'_1$. Then the global price of the composite commodity on the production side may be taken to be equal to the one on the consumption side. If the composite commodity has different compositions on the production side from those on the consumption side, then all those disproportionate elementary commodities are free goods having $p_i = 0$. The distributions of the global prices, $P_1$ and $P'_1$ between non-free elements are, however, indefinite. Non-free $p_i$s may take on any positive set of values as long as $P_1 (= P'_1)$ equals the weighted sum of $p_i$s.

As an extreme case of the above we may point out the case of the composition being the same on both demand and supply sides. We then have $h_1 = h_2 = \ldots = h_{m-1}$ which can be normalised such that $h'_i$s are unity. Then no element becomes free, but the prices of the elements are indefinite. Nevertheless, this state of affairs is on a knife-edge, because if there is an autonomous increase in the demand for one of the elements, then all others become free; the global price $P_1$ is reduced to the price of this element $i, p_i$.

Finally, we are concerned with a particular case satisfying the following conditions:

(a) There are two composite goods *I* and *II* consisting of the set of elementary commodities $1, \ldots m-1$ and the set of $m, \ldots n$, respectively.

(b) The composition of $I$ on the demand side may be different from its composition on the supply side; similarly for $II$.
(c) $I$ and $II$ are gross substitutive on the demand side, while we assume, for the sake of simplicity, that they are independent on the supply side.

Suppose now there is an autonomous increase in the demand for some element $i$ of $I$. It yields a rise of its price $p_i$ and, hence, in the global price $P_I$. Some elements of $I$ may become free, while the prices of the others may take on any non-negative values subject to

$$P_I = p_1 k_1 + \ldots + p_{m-1} k_{m-1} = p_1 k'_1 + \ldots + p_{m-1} k'_{m-1}$$

Consumers will then make a substitution between $I$ and $II$. The demand for $II$ expands as $P_I$ increases. This will give rise to an increase in $P_{II}$. Some elements of $II$ will be free, but others have prices which are indefinite, provided that they satisfy:

$$P_{II} = p_m k_m + \ldots + p_n k_n = p_m k'_m + \ldots + p_n k'_n$$

This increase in $P_{II}$ further creates a substitution between compositions $I$ and $II$, which stimulates the demand for $I$. We then have a further increase in $P_I$ that will give rise to a substitution in favour of $II$ against $I$. This process will continue until $P_I$ and $P_{II}$ settle at their new general equilibrium levels.

## 7

The theory of complementary goods discussed in the previous sections may be translated into the language of the theory of intrinsic complementarity (Morishima, 1959, pp. 188–202); this article is a formulation of a corresponding chapter of Hicks' 1958 book (pp. 166–8). Following Hicks, I think of the consumer as choosing between what he calls 'objectives' of his economic activities and I define the utility function as a function of a number of objectives. It describes the consumer's subjective valuation of them. To attain each objective there are alternative 'means', the relationships between an objective and its means being of a technological nature. A theory pursuing a similar line of thought is later developed and claimed by K. Lancaster (1966, pp. 132–57) as a 'new approach to consumer theory'. He terms objectives 'characteristics' and means 'commodities',[3] and thinks of the utility as a function of characteristics $z$ and of the consumer maximising $u(z)$ subject to the budget equation $px = M$ with the linear or non-linear production functions, $z = F(x)$; in the case of $F(x)$ being linear we have $z = Bx$, with the matrix of production coefficients, $B$, regarding characteristics.

What I have discussed in the previous sections may be stated in the following way. The production of a single commodity by the firm is a joint production of characteristics, because a commodity is interpreted as a bundle of them in fixed proportions. Then, on the demand side, the consumer chooses a collection of commodities which maximises her utility function. In such a case, as we have seen, the prices of characteristics are not certain and remain indefinite, whereas the prices of commodities – what Fanno calls their global prices – are determined. It is impossible, therefore, that the firm attributes, in a clearly defined way, the global cost of production of a single commodity to costs of producing the individual characteristics it contains.

Characteristics are usually offered in combination. In order to obtain a pure characteristic or 'element' we have to apply a further production process of refinement to a gross product of a firm that is usually made available in the form of a hybrid of a number of characteristics, in order to extract a single kind of element from it. Then this extract itself forms a single commodity for which a price is uniquely determined. The bundle of characteristics contained in a single commodity may be evaluated in terms of prices of its elementary characteristics. The total value of characteristics of a single commodity in terms of these prices is usually higher than the price of the commodity as a hybrid. However, if the difference between them is lower than the monetary estimate of the work needed to be made within the household for the 'refinement', then the consumer will buy purified characteristics in the market; otherwise she buys the commodity (say, salmon) in its entirety and discards, at home, the part she does not want to eat. The popularity of DIY shops in recent decades exemplifies the fact that production within the household may be carried out very efficiently.

## 8

Having examined the joint production economy under the assumption of perfect competition, Fanno studies how the economy works if supply is monopolised. He then moves to 'the case of linked markets under a monopoly regime', in which he is concerned with markets of jointly supplied commodities $i$ and $j$ by monopolist $a$ of country $A$ and monopolist $b$ of country $B$. This is referred to by Fanno as a case similar to Marshall's rival supply (Marshall, 1907, p. 390), although the market discussed by the former is one with two rival suppliers (that is, the market of duopoly), while the latter's rivalry means that the commodities supplied by the rival firms are competitive with (or substitutive for) each other. Of course, what Fanno has examined is more complicated and more significant from the viewpoint of application to the real world than the pure theory of duopoly, because Fanno's competitors being different countries, $A$ and $B$, repercussions are not confined within the regime of relative prices but extend to the field of the exchange ratio between the two countries.[4] He himself realises that the nationwide monopoly in each country over an internationally linked market is, in fact, an economy of duopoly, as he states that monopolists of the two countries are 'partial monopolists' in this case.

We may, therefore, put Fanno's analysis of linked markets under monopoly into the form of the Cournot-type or Stackelberg-type model of duopoly. It is then easily seen that it may alternatively be put in terms of the theory of games. We may derive a proposition that the equilibrium Fanno obtains for these linked markets is of a Nash equilibrium type. We may also extend this proposition to the case of oligopoly or that of competition among some limited number of rivals. Contemporary economic theorists will naturally understand that the theory of core follows, so that the general equilibrium solutions are acquired in the limit of the number of rivals tending to be infinitely large. It must be pointed out here that Fanno's work is different from the usual general equilibrium analysis of international trade such as that by Mosak (1944) which entirely ignores the problem of monopoly, duopoly and oligopoly; it is therefore difficult, though not impossible, to obtain from the latter a view that a state of general equilibrium emerges when infinitly many rivals compete each other.

The prices of joint produces are determined under monopoly, in exactly the same manner, as we have observed in the case of perfectly competitive markets. We assume that produces are supplied in the constant proportions, $k_1, k_2, \ldots k_{m-1}$, by the monopolist, whilst consumers buy them in proportions, $k'_1, k'_2, \ldots, k'_{m-1}$. As before, let $h_i = k_i/k'_i, i = 1, \ldots, m-1$, and let $h_1$ be the smallest among $h_i$s. Where the demand $X_i$ does not exceed its supply $Y_i$, we at once have $Y_i > X_i$ if $h_i > h_1$, whilst $Y_i = X_i$ for $i$s such that $h_i = h_1$, whenever $Y_1 = X_1$. This means that all outputs $i$ with $h_i > h_1$ become free, only those outputs $i$ with $h_i = h_1$ can have positive prices. Even all of the latter may become free goods. This is because if there is an exogenous demand for commodity 1 by the amount $a$, so that $Y_1 = X_1 + a$, then we have $Y_j > X_j$ for all $j$.

In addition, we may say that the prices of non-free goods, if they exist, remain indefinite. Only definitely determined is the global price of the joint outputs, $P_1$, but its distribution into the prices of non-free elements is left undetermined, as long as the aggregate of them equals $P_1$. This has been seen for the case of perfect competition above. These rules of pricing for joint outputs are easily extended, without any modification, to the case of duopoly and oligopoly.

It is obvious that the cost of production of the composite commodity $(k_1, k_2, \ldots, k_{m-1})$ is not attributable to those elements which are free. It is distributed only among non-free elements, but the distribution is not unique but uncertain for the reason discussed above. Moreover, all these elements other than commodity 1 easily become free if there is an exogenous demand for commodity 1. The same thing happens if there is an exogenous demand for commodity 2 with $h_2 = h_1$. The structure of the price of non-free commodities is thus very unstable with respect to a shift of demand for elements of the composite output; the allocation of the cost of joint production to the elements swings, suddenly and discontinuously, from one corner of the domain of possible allocations to another. These result will hold true even though we formulate the model of imperfectly competitive markets in the way of game theory.

## 9

Finally we need to make a comment on von Neumann (1945–6) and Sraffa (1960) on joint production in terms of linear processes. Let $A$ and $B$ be input and output matrices. There are $u$ processes and $v$ goods, so that $A$ and $B$ are $u \times v$ matrices. The intensity vector of processes is denoted by $x$ and the price vector by $p$. We are concerned with a stationary equilibrium; thus von Neumann's $\alpha$ and $\beta$ are set to equal 1 as the rate of profits and the rate of growth are assumed to be 0. We then have equilibrium inequalities:

$$Bx \geq Ax, \tag{10}$$

$$pB \leq pA \tag{11}$$

If in (10) > applies to good $j$, then $p_j = 0$ (the rule of free goods); if in (11) < applies to process $i$, then $x_i = 0$ (the rule of profitability). Von Neumann has shown that there is a non-negative, non-zero set $(x, p)$ satisfying (10) and (11).

Suppose now, as before, that the first $m - 1$ goods are jointly produced by the first process and no other process can produce any of the first $m - 1$ goods. Then B takes the form:

$$B = \left[\begin{array}{c|c} b_{11} & \\ b_{21} & 0 \\ \cdot & \\ \cdot & \\ \cdot & \\ b_{m-11} & \\ \hline & \\ 0 & C \\ & \end{array}\right] \tag{12}$$

where $b_{11}, \ldots, b_{m-11}$ are positive and $C$ is a $(u - 1) \times (v - (m - 1))$ submatrix. $A$ is similarly partitioned:

$$A = \left[\begin{array}{c|c} a_{11} & \\ a_{21} & \\ \cdot & D \\ \cdot & \\ \cdot & \\ a_{m-11} & \\ \hline & \\ 0 & E \\ & \end{array}\right] \tag{13}$$

This partition of matrix $A$ implies that commodities $1, \ldots, m - 1$ are jointly consumed. Our analysis in the following corresponds to the previous Neo-classical analysis of the case where goods $1, \ldots, m - 1$ are jointly produced and jointly consumed.

Let us now assume that commodity 1 is the main product in the sense that $b_{11}/a_{11} = \max(b_{11}/a_{11}, \ldots, b_{m-11}/a_{m-11})$ and process 1 is productive with respect to each component commodity – i.e. $b_{i1}/a_{i1} > 1$ for all $i = 1, \ldots, m - 1$. Suppose there is an increase in the demand for commodity 1. Then:

$$Bx' \geq Ax' + (a, 0, \ldots, 0)^{\mathrm{T}} \tag{10'}$$

$$p'B \leq p'A \tag{11'}$$

where T stands for the transposition of the vector to which it is applied. From (12) and (13) we see that (10′) holds at an $x'$ such that $x'_1 > x_1$ and $x'_i = x_i$, for $i = 2, \ldots v$, where $x$ is an output solution to (10). Then (10′) holds with strict inequality for commodities $2, \ldots, m-1$, so that they are all free. This is exactly what we have seen in Section 6 above.

This is nothing else than a mere exercise applied to the von Neumann model. Sraffa (1960), however, entirely neglects the demand–supply subsystem (10) and concentrates his attention upon the proceeds–cost subsystem (11). Moreover, he replaces inequality $\leq$ of (11) by strict equality. This modification is serious. First of all, the modified (11) does not necessarily have a non-negative (non-zero, of course) set of prices. Secondly, Sraffa misses the free-goods phenomena that are intrinsically involved in the joint production problem, by neglecting the demand–supply equilibrium conditions of the inequality form (10). Sraffa is half von Neumann; the half he misses out is most essential in dealing with the problem of joint production. Fanno is more orthodox than him.[5] This is the reason why I respect von Neumann but am critical of Sraffa.

*Note*: At the time when Fanno wrote the material contained in this volume, it was not necessarily the care that full reference details were provided. It has proved impossible to make the Bibliography section complete.

# Introduction

**1.** Thanks to the work of Cournot,[1] Jevons,[2] Edgeworth,[3] Marshall,[4] Auspitz and Lieben[5] and many others the theory of prices has made notable advances. The one-sided views epitomised in the Classical School's principle of the cost of production on the one hand and the Austrian School's final level of utility on the other are giving way to a broader and more complex vision in which these two principles, synthesised in two curves of demand and supply, are coordinated and fuse together.[6] And with this harmonious fusion of principles that were long regarded as discordant and irreconcilable the theory of prices is gaining greatly in precision and rigour.

This is not to say that the theory is approaching perfection. As is clear from the works of the authors referred to above, what we have is a first approximation to reality rather than a complete theory. In fact the economists just mentioned, together with their disciples, treat the two curves of demand and supply that combine to form each single price as the initial data of the problem – their starting-point in the quest for the laws of prices, which they represent as a process of partial independent equilibria in which the price of each individual good or service is determined independently of the prices of all the others.

It is true that in developing and extending Mill's brief studies of 'a few special cases of value'[7] these economists also attempt to study a few groups of connected prices, thereby showing some recognition of the existence of a certain interdependence between particular categories of prices, but these supplementary studies are accorded far too little importance to be able to modify the fundamental lines of those theories. Thus cases of independent prices always remain the rule for these economists and cases of connected prices an absolute exception.

But such a way of seeing things does not precisely correspond to reality. It is true that each individual price is established as a function of the curves of demand and supply, but the latter in turn result from a combination of factors. For example demand for a particular good is influenced by the price of other goods that are demanded with it or instead of it as complementary or substitute goods while supply of each good is influenced by the price of all the goods and services that combine to make it and that constitute elements in its cost of production. Therefore every curve of demand and supply (if we can still speak of curves) is the result of the combined action of various prices, and since all the individual examples of demand and supply curves combine to form each individual price, in the final analysis each price is determined as a function of all the others.[8] Prices of all goods and services, far from being a product of myriad partial independent equilibria, are in fact the product of a single complex system of general equilibrium.

This connection between prices has not been overlooked by scholars – with typical perspicacity Jevons himself[9] asserted that cases of joint cost, generally considered the exception, are in fact the rule; but he did not therefore simply abandon the study of partial equilibria in favour of the study of general equilibrium, perhaps because a more complete elaboration of the former was necessary before science could arrive at a synthetic vision of the latter; for in fact it is only when the theory of partial equilibria reaches the highest level of perfection that it gives way to more general conceptions. A successful attempt at a synthesis was made by Irving Fisher in an early work.[10] But it was only with the Lausanne School that this new direction became the rule. With the recognition of the close interdependence of all economic and social phenomena – something that then became a fundamental canon of our discipline – this school (as personified by Walras and Pareto[11]) replaced partial analyses of special groups of phenomena with a general synthetic vision by means of which it managed to present the whole economic system as a real coordinated whole and reconstruct the theory of prices on the basis of their close connection.[12] With this renewal the theory of prices reached the final stage of its journey, and despite the occasional apparent deviation it has broadly speaking continued to follow this same line of development. From the vague one-sided principles of the Classical and Austrian Schools to the rigorous formulation of partial equilibria by Jevons and Marshall and from this latter to the synthetic conception of the Lausanne School, it has gone through successive stages of which each formed the necessary preparation for the one that followed; thus the series of theoretical contributions made by the various schools on the way has apparently culminated in the final goal with the theory of general equilibrium.

But although the Lausanne School may have had the great merit of raising the theory of prices to the highest level of scientific rigour, it still did not manage to bring that theory to full development, for it has provided us with an outline of the laws of prices rather than with the complete picture. In fact to know these laws thoroughly it is not enough to state that all prices are interconnected nor is it enough to provide the complete system of equations simultaneously determining the prices of all goods and services. We also need to show the nature of the relations between the different prices, the direction of their respective movements, the relative strength of these movements, and the way in which the various groups of prices are linked. But the Lausanne School is silent on all this, so despite all its innovative and fruitful research the field of the correlations between prices remains for the most part more or less virgin territory with potential for very extensive development (as Professor Pantaleone recently reminded us).

**2.** The research I have undertaken in the present work is intended to fill this lacuna. The problem is that correlations between prices are so vast and complex that they can never be studied *en bloc* in their totality. Minute analysis is required to trace every twist and turn of the guiding thread by which each individual price is connected to all the others, and this cannot be done except by breaking down the whole system of correlations between prices into groups or categories, and proceeding to the study of each group individually. Therefore even though the complete law of prices can only emerge from an overall view it is nonetheless a methodological necessity that we spend some time on a study of partial analyses before we can arrive at such a view.

What are these groups into which the full system of correlations between prices can be broken down, and what are the groups of correlations to be studied? The way that prices communicate their respective movements to each other is unique to each case. Since each price is fixed as a function of demand and supply the action of the price of each individual good on the prices of other goods cannot be explained except through the demand for and supply of these other goods. The whole field of correlations between prices can therefore be broken down into as many groups as there are different ways in which levels of demand for and supply of the various goods are connected.

It is supply that functions *par excellence* as a connecting link between prices whenever there is a relation between the production of two or more goods such that a change in the production of one good modifies the production and therefore the supply of the other good or goods. In fact whenever a relation of this kind exists, a change in the price of any good, while modifying the supply of that good, reacts through this on the supply of the other goods, increasing or reducing that supply and therefore raising or lowering the price of those goods depending on whether the increase in production of the first good is matched by a corresponding increase in the production of the others or by a fall in that production. The first alternative occurs whenever the goods in question are made by the same production process; the second alternative occurs whenever they are made by different production processes so that expansion of the production of one takes away from the others a part of the initially available coefficients of production and reduces that production. Thus we have two groups of correlations in respect of supply: correlations between prices of goods produced jointly and correlations between prices of goods produced in competition.

As for demand, to begin with it is well known that each curve of demand is only the transformation of a corresponding curve of utility.[13] Now far from being autonomous and depending only on the quantity of the good in question available, the marginal utility of each good also depends on the quantity of all the other available goods.[14] Hence while a change in the quantity of a given good modifies its marginal utility, such a change also modifies the utility curve and hence the demand curve and the price of all other goods. Thus we can say that the prices of all goods are connected through demand.

There are two principal kinds of relation between different prices established by demand. To begin with there are complementary goods – goods that provide directly for the satisfaction of needs only inasmuch as they are associated with each other, or that provide better for the satisfaction of needs jointly than they do separately.[15] Where these goods are concerned increase in demand for one necessarily brings about an increase in demand for the other, and the level of complementarity common to an infinity of goods therefore establishes an initial order of relationships of interdependence between prices.

Other goods however compete with respect to demand – goods that are able directly to satisfy the same needs and that can therefore be employed in place of one another. To this category belong goods known as substitutes, or goods that are susceptible to economic reproduction in the sense put forward by Ferrara.[16] Where such goods are concerned an increase in demand for one takes place at the expense of demand for the other, and the level of substitutability therefore establishes the second order of relations of interdependence between prices.

These two groups of complementary and substitute goods include direct goods and instrumental goods and hence comprise all categories of goods, and the internal relations between the goods in these groups include all the kinds of relations that are established between the prices of different goods by demand.

To summarise then, there are four fundamental groups into which the system of correlations between prices may be broken down, as follows:

(1) correlations between prices of goods produced jointly;
(2) correlations between prices of goods produced in competition;
(3) correlations between prices of goods that are complementary with respect to demand;
(4) correlations between prices of goods that compete to meet demand.

I shall be studying each of these four groups separately, looking at how price equilibrium is established within each group and at how each individual price acts and reacts on others in the group. This investigation will lead us to the laws of simple correlations between prices for individual groups, and these laws will form the basis of my thesis in Chapter 1 of the present work.

**3.** These four groups into which the entire price system may be broken down for purposes of study are not however in reality four closed and isolated groups – as might seem to be the case in view of the discussion in Chapter 1 of this work. They arise from the classification of goods into complementary and substitute, goods made by joint production and goods produced in competition, a classification which no doubt reflects real objective differences, but like all classifications it follows the kind of clear and precise qualitative dividing lines that simply do not exist as such in the real world. In fact no good actually belongs exclusively to any single one of these categories, for every good is complementary to another and at the same time a substitute for yet another or others. Each good is produced jointly with some goods and at the same time in competition with others, so that while it has a foothold in each group it equally belongs in all the groups at once. It is just this singular ubiquity proper to each good that forms the link between the different groups.

To the extent that each good belongs to a given group it is governed by the laws of that group, but insofar as it belongs to other groups it is governed by their laws too. Far from being affected purely by the actions and reactions of the prices in one particular group, the price of any good is actually affected by actions and reactions in all the groups, and far from being limited purely to individual groups, correlations between prices are actually complex correlations involving all the groups at once. So the movement of any price in the face of any sort of disturbance is not governed by the laws of equilibrium of the prices in any particular one of the groups under consideration, but results in fact from the theoretical individual movements of each price – in different directions and with different force – within the limits of each of the four groups to which it simultaneously belongs.

If these really are the relations between prices, laws arising from the study of individual groups in respect of them cannot of course be anything other than imperfect and incomplete. So a theory of prices that aspires as far as possible to mirror the actual state of affairs cannot restrict itself to these initial researches nor can it accept their conclusions as definitive. It needs to push on and develop a more sophisticated picture, and the essential elements of this will be provided by the same analytic research. If it is true that the complex correlations are only the result of the simultaneous combined action of simple individual correlations, it follows that to construct the laws of the former it is necessary to coordinate, link together and fuse the laws of the latter. Therefore having broken the price system down into individual groups of correlations and having studied the laws of these groups, we shall then have to abandon the logical device of subdivision by groups and invert the initial process so as to reconstruct the system as a single whole. To this end we shall have to take the individual groups one by one and bring them together, coordinate them, fuse them, and see how the correlations between prices in each group progressively act and react on the prices in the other groups. It will only be after this work of synthesis, coordination, and fusion that with the artificial divisions into groups gradually eliminated the complete structure of correlations between prices will finally become apparent in its real shape as a vast, complex, tightly-knit web of relations holding all prices together by a common chain of solidarity.

**4.** But together with this mass of correlations deriving from the relations between production and consumption, i.e. together with the interweaving of individual demand and supply curves, other types of relation hold between prices and we shall also have to concern ourselves with these. I am referring to relations of a monetary nature.

In advanced economies exchanges take place through the direct or indirect agency of money. Assessments of the value of all goods are made with reference to the monetary unit. The prices taken into consideration are monetary prices. The modifications of prices that are registered are measured on the basis of changes in their monetary expression, and all this process of the formation and transformation of prices goes on with the aid of a vast monetary machinery. Therefore if it is to be complete, a theory of prices, as well as highlighting the network of reciprocal relations between prices, must show in what way the whole system of correlated prices functions through the monetary machinery; in what way prices are connected momentarily; and in what manner, in what direction and to what extent all other prices move simultaneously in reaction to variations in one money price. And to be as close as possible to reality such a theory must fit the laws of monetary circulation in with the laws of the complex correlations between prices and create an overall synthesis.

A few reflections will be sufficient to make plain the need for this. I shall not here undertake a further study of the conditions governing monetary equilibrium, which I discussed at length in an earlier work to which the reader is referred.[17] It is sufficient to note that given a determinate quantity of money in a market forming the basis for a determinate development of credit, and given a determinate velocity of circulation of the two parts into which the overall circulating medium is divided, monetary equilibrium is established at the price level to which corresponds a total value of the mass of circulating goods precisely equal to the mass of the circulating medium (multiplied by the respective velocities of circulation) available on the money market at a discount rate equal to the interest rate.[18] The prices of individual goods together form this level of prices, and according to a well known law in equilibrium conditions these prices are proportional to the level of marginal utility (Pareto's elementary ophelimity)[19] of the respective goods. Now it is evident that if the level of marginal utility of a good changes, the price of that good will change with it, but at the same time because of the interdependence already noted the levels of marginal utility of the other goods change too. Some of them rise, others fall, and the prices of the respective goods tend to rise and fall correspondingly. This general rearrangement of individual prices gives rise to a change in the overall value of the mass of circulating goods.[20] But this change, occurring while the other conditions of monetary equilibrium remain unaltered so that the mass of the circulating medium corresponding to the normal discount rate remains unchanged, acts directly on this rate and pushes it away from its normal level. Now except in disturbances of a temporary nature the deviation of the discount rate from the interest rate creates a disequilibrium that finds its own correctives in the price movements which it tends to provoke. A fall in the discount rate tends to raise prices while a tension in the rate tends to lower them. So in the face of a shift in the discount rate away from its normal level produced by the rearrangement to which prices are initially subjected to fit the new marginal levels of utility, prices tend to move together *en bloc* in one direction or the other; and they continue to move until as a result of their movement the value of the circulating goods again comes to be equal to the mass of the circulating medium corresponding to the normal discount. Once discount and interest rates again coincide at this level, the market returns to an equilibrium position where individual prices, irrespective of their new individual levels, are graduated proportionately to the new marginal levels of utility of the respective goods.[21]

Now in the transition from one equilibrium position to another the prices of all goods change without exception, and the shift in each individual price is represented by the difference between its respective levels in the two equilibrium positions. It is this shift – the effective movement of each individual price – that represents the ultimate actual tangible result of all the actions and reactions of any price when subjected to any kind of disturbance. Applied to all prices simultaneously this result sums up the general law of prices, which is exactly what we are looking for.

These shifts in individual prices are not however for the most part proportional to the shifts in the levels of utility of the respective goods. For some goods they are greater, for others they are smaller, depending on the case. And this is natural, for in their movements individual prices must fit both the requirements of the hedonistic equilibrium and those of monetary equilibrium and they cannot be exclusively subjected to the laws of either.[22] Therefore the shift in every other price and all prices together in the face of a change in a single price is the ultimate result of the combined action of complex correlations between prices and the laws of monetary equilibrium. The mere recognition of this fact is enough on its own to clarify and confirm the necessity of integrating the study of complex correlations between prices with the study of the laws of monetary equilibrium.

**5.** I have now set out the broad outlines of this work and given an idea of its structure. In a first approximation I shall enumerate the groups of correlations and study them one by one; I shall examine how price equilibrium is established within each group and how each individual price acts and reacts on the others in the group; and I shall demonstrate the laws of the partial correlations that emerge.

After this in a second approximation I shall move on from analytic research by group to a more general study that will integrate, coordinate and fuse the results of the first. To this end I will bring together the various groups considered separately in the first approximation and look at how the correlations between the prices in each group act and react progressively on the prices in the other groups. The combined effects of all this work of analysis and synthesis will be to produce a complete framework of the complex correlations between prices. Finally in a third approximation I will coordinate the laws of the complex correlations between prices with the laws of monetary equilibrium; and by showing how the system of correlated prices acts and functions thanks to the activity of the monetary mechanisms that form its complement I will ultimately be able to reconstruct the complete theory of prices.

Clearly such a work is going to be vast, arduous, and complex, and interwoven throughout with minute analyses and partial general syntheses. Each page presents a new problem, the solution of each new problem prepares the way for the solution of subsequent problems, and the whole study unfolds through a succession of rigorous demonstrations, chains of reasoning often fraught with difficulties that even the most skilled reader will have difficulty getting through without a struggle.

I plan to develop this work step by step in separate monographs, and later if the opportunity arises I will bring them together in a single volume. Keeping closely to my programme I shall begin with the analytic study of the groups of correlations mentioned, taking as my starting point the group of goods produced jointly. This group, whose importance I hardly need to emphasise, will be the basis of this first monograph.

# 1 Preliminary Observations on Supply at Joint Cost

**1.** Before embarking on this study we must consider what is to be understood by the term 'joint cost' since economists are not in complete agreement on this point. They waver between two extreme concepts, one very broad and the other very narrow. In a narrow sense they take supply at joint cost to mean the supply of two or more goods or services that are the product of a single production process so that an increase in the production of one cannot take place without a simultaneous increase in the production of the others.[1] In a broad sense they take it to refer to any supply of two or more goods or services whose production arises directly from the same production process.[2] In this latter sense supply at joint cost includes both supply of goods or services that can take place only connectedly and supply of goods or services which could be produced separately but which are in practice produced by one single process for reasons of convenience or economy; in other words it includes not just cases coming within the narrow concept but many other cases as well.

This excessive extension of the concept of joint cost does not seem to me to be either correct or scientifically rigorous.[3] To have scientific value and importance in fact, any group of phenomena arising from any kind of classification must include phenomena that are homogeneous not merely in form but also in substance. Now this is very far from being the case with the group of supplies at joint cost in the broad sense referred to above. There is in fact a big difference between production of several goods or services carried out at shared cost because it is not possible to do otherwise and production carried out at shared cost for contingent reasons of convenience or economy. The one occurs spontaneously, the other purely as a result of considerations of convenience, the one implies close and indissoluble relations between the prices of the joint goods while the other implies merely contingent and temporary relations. There is a difference between the nature of the correlations between prices in the two cases, there is a difference in the way prices behave and a difference between their laws. In sum we are talking about two categories of phenomena that are substantially different despite their seeming similarities; two categories that must not be confused and that need to be studied separately. Thus in this study I have jettisoned the broader concept of supply at joint cost as being too vague and indeterminate and I will treat the two categories of phenomena covered by it as two distinct groups, assigning different names to each of them. I will use the name of supply at *associated costs* or in Pigou's terminology, *common costs*,[4] for those in the second group, since it seems that the terms 'associated' or 'common' are appropriate to the condition of a contingent, occasional, or temporary union of costs characteristic of this group. I will use the term *supply at joint cost* solely for that supply where the union of costs is constant, necessary, and indissoluble. In other words I shall go for the narrower concept and limit discussion in the present work to cases that fit into it.

Since J. S. Mill[5] many economists have studied production at joint cost on the basis of this concept – until recently the dominant one. Virtually every handbook or study of economics devotes a chapter or at least a few paragraphs to the topic.[6] But few authors do anything more than textually reproduce or paraphrase Mill's classic pages, adding nothing substantially new. Even those who, like Marshall for example,[7] give rather more thought to the topic, can find only a partial solution to it. They limit themselves to studying the very special case of free competition in a closed market and even this they do only very incompletely. They make no effort to provide anything like a comprehensive study of all possible cases.[8] Thus the theory of supply at joint cost remains still largely incomplete, and it will be the aim of this monograph to fill in the lacunae that remain in it.

**2.** To this end I shall study supply at joint cost both in a free market and under a monopoly, and I will present this study as two successive approximations: the first in relation to the conditions of a closed market, the second in relation to the conditions of linked markets.[9] I shall study in what way and according to what laws production equilibrium is established in the different instances; what are the correlations between prices of goods having joint costs, and in what way and through what price movements equilibrium is reestablished once it has been disturbed. Finally I will examine the way in which individual demands for and supplies of the goods in question behave in relation to each other in different instances and how – thanks to their reciprocal adaptation – they combine to restore equilibrium.

My strategy in the first approximation will be to move directly to a consideration of the curves of market demand and supply, regarded as separate from the corresponding curves of utility, taking account of the marginal utility of money – which we shall suppose to be known and constant (a hypothesis which we shall see to be legitimate though not strictly rigorous in a closed market). In the second approximation – where I shall be dealing with linked markets – it will however no longer be possible to regard the marginal utility of money as constant, even in a broad approximation. In this case therefore I shall have to operate with utility curves and present the law of demand and supply as complex functions with several variables.

The plane curves of demand that I use in the first approximation normally have almost the same trajectory, and differences between one curve and another are basically differences of degree. If all other conditions are the same (a hypothesis that is implict in the nature of a plane curve having for analytic expression a function of a variable) the quantity demanded will generally tend to decrease with the rise in the price; or, to put it another way, purchase prices tend to go down where there is an increase in the quantity available. This represents a kind of general law, though one that does admit of some exceptions.[10]

The relation between the decrease in demand for a good and the rise in the price of that good (assuming that all other conditions remain unchanged) is called the elasticity of demand; this may also be regarded as the relation between the fall in the demand price and the increase in the quantity available if we take the quantity as the variable and the price as the function. Demand is said to be the more elastic to the extent that the price falls less rapidly with the increase in the quantity; or, to put it another way, the greater the force of the fall in the quantity demanded with the rise in price. If the demand price varies exactly inversely with the quantity, the overall monetary value of the quantity acquired or the sum expended in this acquisition remains constant for any price and quantity. The corresponding curve of demand (with quantities as abscissae and prices as coordinates) is an equilateral hyperbole; and the elasticity of demand individualised by a curve of this kind is conventionally considered by economists to be equal to one. If the curve of demand constructed in the way indicated is less steep than an equilateral hyperbole, the overall value of the quantity acquired rises with the increase in the quantity or falls with the rise in the price.

In this case the elasticity of demand is said to have a value higher than one.

If on the other hand the curve of demand is steeper than an equilateral hyperbole, the overall value of the quantity acquired decreases with the increase in the quantity or increases with the rise in the price. In this case demand is said to have an elasticity lower than one.[11]

It seemed the right moment to discuss this now, since later I shall need to refer fairly frequently to the elasticity of demand and its level of intensity. But in this connection it will help to note that elasticity of demand quite often varies markedly from one point to another on the curve, unless the curve is a demand curve that follows the trajectory of an equilateral hyperbole throughout, in which case elasticity of demand is equal to one at every point. When we speak of elasticity of demand it is therefore a good thing to specify the point on this curve to which we are referring in order to avoid ambiguity or possible mistakes. Consequently whenever I speak of elasticity of demand I shall mean elasticity relative to the point or zone of such equilibria as are being discussed even if I do not mention that point or zone explicitly.[12]

**3.** In the first approximation supply will be represented by a curve of monetary costs which we shall suppose to be obtained by the transformation of an initial curve of costs; these will be taken, with due reference to marginal principles, in the Cairnesian sense of physio–psychological costs,[13] i.e. the sum of sacrifices made by individual industrial groups[14] in carrying out the work of production.

These curves mark out the trajectory of unitary monetary costs as a function of the quantities produced, and they are known by the name of *curves of costs* or *of supply*. But from now on in order to avoid any ambiguity it should be noted that the curves I shall be studying are curves of supply relative to the whole market and not to one individual company,[15] so they are an expression of the cost that the production of a given quantity of a determinate good incurs for each metrical unit in a given market and time.[16] But some explanation is needed here of the significance of unitary cost and of the way of calculating it. The case of a free market must be distinguished from that of a monopoly. In the first case there are several competing firms, each of which incurs a different cost by reason of its particular technical and economic structure. The unitary cost for each firm is calculated by dividing the total cost by the quantity produced, which yields the average unitary cost. The average unitary costs of the various competing companies arranged in order of magnitude form the scale of costs. The regulatory market cost is given by the average unitary cost of the marginal company whose supply needs to be boosted to have a total supply equal to the demand at the price corresponding to this cost. If we imagine an infinite number of companies so that differences between respective average unitary costs are infinitesimal, the scale of costs becomes a continuous function of the quantities brought to the market, and it gives rise precisely to a curve of supply that is simply the curve of average unitary costs of successive marginal companies or, in short, the curve of *successive marginal costs on the market*. This is the curve I shall use in cases of free competition, and it may be one of rising, uniform or falling costs. But we should immediately note that only in cases of falling costs and exceptionally of uniform costs are true equilibrium positions found in a free market. In fact where there are falling costs, just as the price is dictated by the marginal cost that in this case is the lowest, so companies with higher costs are sooner or later going to disappear. Their disappearance changes the scale of market costs and stimulates marginal companies to expand by further pushing down their costs. So the law of decreasing cost fundamentally mirrors a dynamic state of the market and unstable rather than stable equilibria.[17] In dealing with cases of free competition I will take into consideration only conditions of rising costs and where necessary of uniform costs.

Moving on to cases of monopoly I shall make a small modification in the way of constructing the curve of supply, and the reason for this will be explained very shortly. Here too I shall consider the curve of market supply, but as we shall be dealing with a monopoly regime market supply will mean supply by the single monopolistic company. This supply will therefore be considered *en bloc* and unitary costs will also be calculated *en bloc* by dividing total cost by total production. Where production increases total cost rises and average unitary cost will or will not rise depending on the behaviour of marginal cost – in other words depending on whether marginal cost is higher or lower than average cost and depending on whether the said marginal cost is constant, rising, or falling.[18] This last condition depends on the *point reached* in the development and life of the company.[19] Consequently in correspondence with the extent of its production the company's average unitary cost may be constant, rising, or falling. The succession of these unitary costs constitutes the company's curve of average unitary costs, and in the case of a monopoly this is the same as the curve of market costs. The average unitary cost is the same as the average of the successive marginal costs. Hence the curve of average unitary costs derives from the curve of marginal costs.[20]

In dealing with cases of free competition I shall follow this second curve and in dealing with cases of monopoly I shall follow the first. The reason is this. In cases of monopoly there is almost always a need to deal with the total cost of production, and given the curve of successive average unitary costs it is easy to calculate this total cost. All we need to do is to multiply the unitary cost by the quantity corresponding to it. It is also possible to obtain the total cost from the curve of successive marginal costs, but the calculation involved is more difficult. The total cost is equal to the definite integral of the function of the marginal costs for values of the variable falling between zero and the quantity produced. To avoid the use of integration symbols I shall in cases of monopoly use the curve of average unitary costs instead of that of marginal costs. The reader will now be able to recognise what curves are being referred to on each occasion when curves of costs are being discussed in cases of free competition or of a monopoly, and the two curves will be designated by different symbols. So I can now move directly to the study of supply at joint cost in the two approximations specified above.

# 2 The Theory of Supply at Joint Cost: The Case of a Closed Market

## I IN A FREE MARKET

**1.** To study supply at joint cost requires massive and complex research. To facilitate the task I shall initially concern myself with the simple case of just two joint goods; afterwards I shall move on to more complex cases and then finally come to some general conclusions. Cases of jointly produced goods are very numerous: gas and coke, straw and grain, wool and meat or sheep's milk are the classic examples. I will refer to these concrete examples from time to time for purposes of greater clarity, but in general I will designate goods by letters. Let *A* and *B* be the two goods in question. In a free market the price of goods tends to be equal to the marginal cost of production.[1] Goods produced jointly are also governed by this law, with the difference firstly that prices of goods are two or more than two rather than just one, and secondly that since there is no specific cost for each product it is the global unitary cost that counts. Prices of finished products are based on the unit of measure of these products while the global cost is based on the unit of global production, which may be of the same kind as the unit of measure of finished products, but which may also be different. However since yield prices are set against cost they need to be reduced to their common denominator if they are to be comparable to the latter. Hence if on the one hand we take account of the global marginal unitary cost, on the other hand in calculating the unitary return to which the cost must be equal, account will have to be taken not of the whole price of each unit of the finished product, but of that fraction of the price of each good that corresponds to the fraction of a unit of each good that is obtained from each unit of global production. If we add these fractions of prices together we obtain the returns corresponding to each unit of global production. For purposes of brevity I shall from now on call this return the *global price*.[2]

It now becomes clear how the well known law of the adaptation of price to cost is to be understood in relation to the specific case of joint products. Understood in the manner indicated this law may be stated in the following way: given two or more goods produced jointly, in a free market their global unitary price will tend to be equal to their global marginal unitary cost.

But the goods under discussion, precisely because produced jointly, may for technical reasons be produced in a given ratio to one another. This ratio we shall suppose for the moment to be constant, but at a later stage in the elaboration of my theory I shall treat it as variable. There are then two concurrent conditions for production equilibrium at joint cost; equivalence of the global price to the global marginal production cost and invariability of the ratio between production levels of the various goods. Let $x$ be the quantity of good $A$ obtained by a given global production process and $y$ the quantity of good $B$; let $f(x)$ be the law of demand for $A$ or in other words the price of $A$ expressed as a function of its quantity, and let $\phi(y)$ be the law of demand for $B$; let $k$ be the constant ratio between production levels of the two goods $A$ and $B$; let $F(x+y)$ be the analytic expression of the curve of supply for $A$ and $B$ expressed as a function of the quantity of global production.[3]

This will give us the following two equations:

$$\frac{x}{x+y}f(x) + \frac{y}{x+y}\phi(y) = F(x+y) \tag{1}$$

$$\frac{x}{y} = k \tag{2}$$

These two equations contain only two unknowns overall so they are sufficient to solve the problem, and the equilibrium is perfectly determinate. Putting into words the meaning of these equations I would say that where two or more goods are produced jointly equilibrium is established at prices for the individual goods such that they determine a global unitary price equal to the global marginal unitary cost of production, and that in relation to the laws of demand for each good they cause a demand for each such that it stands in the same relation to demand for the other good or goods as the quantities of the various goods produced jointly stand in between themselves.[4]

If we study the two equations we see that equilibrium is established at given prices as a function of given quantities. But prices, quantities, and costs are correlated and depend respectively on the form of the functions $f(x)$, $\phi(y)$, $F(x+y)$ and on the value of the constant $k$. Equilibrium is established at these given prices in correspondence with these given quantities only on condition that these functions have this given form. A change in one of them or a disturbance of any other kind (like a tax or a premium on production) is enough to shift the initial equilibrium and move the market to a new equilibrium position. During the transition from the one to the other the prices of the two goods $A$ and $B$ change, but they are connected in their movements and indeed there is a close correlation between them which is established through the relations between the various factors in the equilibrium. So it is necessary to examine and study the nature of this correlation. We shall therefore consider a given equilibrium position and suppose that the initial conditions of it change consecutively one by one. The disturbances we shall study in turn are as follows:

(a) a change in demand for $A$;
(b) a change in demand for $B$;
(c) a change in the curve of global costs;
(d) the introduction of a tax;
(e) the introduction of a premium on production;
(f), (g) a change in the ratio between the production levels of the two goods, i.e. a change in the constant $k$.

We shall study the effects of all these disturbances, i.e. their impact on the equilibrium and particularly with regard to absolute and relative price movements.

**2.** (a) First of all then let us suppose that there is a *change in demand for good A*. A change of this kind resolves into a change in the form of function $f(x)$ – from now on, corresponding to the successive quantities $x$ there are demand prices that are different from – i.e. greater or less than – the previous ones.[5] Let $f(x)$ change in such a way that in correspondence with the same price as initially there is demand for a greater quantity, in other words suppose what is commonly but inaccurately called an increase in demand for $A$.[6] The price of $A$ will rise and this will produce an increase in production of $A$ and a proportional increase in production of $B$, but as production of $B$ increases, i.e. as the value of $y$ increases, while the form of $\phi(y)$ remains unchanged the price of $B$ will go down. Hence through disturbances resulting from variations in demand for $A$ the price of $B$ tends to change in the opposite direction to the price of $A$.

So far my conclusions coincide with those arrived at by others who have studied the question, but now I want to go much deeper into these correlations and see in what proportions and according to what laws the price of $A$ and inversely that of $B$ change in the face of a change in demand for $A$.

Various factors affect shifts in the two prices:

I. *The extent of variations in demand for A*. If variations in demand for $A$ are significant then variations in the price of $A$ will also be significant and so will the inverse variations in the price of $B$. But if variations in demand for $A$ are limited, with all other conditions the same the shifts in the two prices will be insignificant.

II. *The elasticity of the new curve of demand for A in proximity to the new price corresponding to the initial quantity*. If within the limits indicated demand for $A$ is very elastic the shift in the prices of $A$ will be small, likewise the increase in the production of the two goods (again relative to a determinate curve of costs) and likewise the inverse shift in the price of $B$. But if demand for $A$ is not very elastic the shifts in the price of $A$, in global production and in the price of $B$ will be substantial.

III. *The elasticity of demand for B*. If demand for $B$ is very elastic, a given increase in global production and therefore in the production of $B$ will be matched by a tiny fall in its price. But if demand for $B$ is not very elastic a given change in the production of $B$ will produce a corresponding movement in its price that is substantial. Now if all other conditions remain the same the more the price of $B$ moves in a given direction, the greater must be the force of the movement of the price of $A$ in the opposite direction if the global marginal return is again to coincide with the global marginal cost.

Hence if all other conditions remain unchanged the shift of the two prices in opposite directions which takes place as a result of a change in demand for $A$ is the smaller as elasticity in demand for $B$ is the greater.

There is however another condition that affects the definitive variations in the prices of $A$ and $B$, namely:

IV. *The trajectory of the curve of marginal costs, i.e. the form of function* $F(x+y)$. An increase in demand for $A$ will produce a rise in the price of $A$, an increase in global production, and a fall in the price of $B$. But if all other conditions remain unchanged the fall in the price of $B$ and the rise in the price of $A$ will be more or less substantial depending on the trajectory of the curve of global marginal costs. Just as global unitary prices must coincide in the successive equilibrium positions with the global marginal cost, so if the latter has a rapidly rising trajectory an increase in demand for $A$ will produce an increase in global production, raise the price of $A$ and lower the price of *B*; but the new prices and the new global costs will soon coincide. Equilibrium will thus be reestablished with a small increase in production and a tiny movement in the two prices. If however the curve of costs rises slowly or follows a horizontal trajectory, a rise in the price of *A* will have to be met by a substantial fall in the price of *B*, if the global price is once again to coincide with the global cost. For *B* to fall substantially in price it is necessary for production of *B* – i.e. global production – to expand considerably. Hence in this case the disturbance in question will be accompanied by a major increase in production and a substantial shift of the two prices in opposite directions.

The conclusions reached so far are also valid (with a few variations) for cases in which there are more than two joint goods involved. Here once more an increase in demand for *A* raises its price and produces an increase in global production with a consequent fall in the price of the other products. But in these cases the rise in the price of a product, i.e. good *A*, produces a corresponding fall not just in the price of one single good but in that of several goods at the same time. This simultaneous fall in the price of several goods has a considerable influence on the form of the new equilibrium. If we exclude the hypothesis of falling costs, global marginal costs in the new equilibrium position cannot but be equal to or higher than the initial ones and the new global prices must be equivalent to these identical or higher costs. Now since for the rise in the price of *A* there is in response to the increase in production of each good a corresponding fall in the price not just of one single good but of several goods at once, the adequation of the new global price to the new global cost will be achieved through a fall in each individual price that is less than in the case of just two joint goods, and through a smaller increase in the production of each good.

If this conclusion is valid in the case of any number of goods produced jointly in comparison with the case of just two goods, it is also of course valid in the case of any number of goods produced jointly in comparison with the case of any other number that is lower than itself.

Hence to summarise we may conclude that given two or more goods produced jointly, when there is a determinate change in demand for one of them, the resulting disturbance will be the smaller – i.e. a new equilibrium will be reached through a change in production of each good that is the smaller and with a shift in individual prices that is the smaller –

(*I*) the greater the elasticity of demand for the good for which demand has changed;
(*II*) the greater the elasticity of demand for the other goods;
(*III*) the more closely costs conform to the law of rising costs;
(*IV*) the greater the number of goods produced jointly. This final condition corresponds to the well known principle of mechanics, according to which the more diffuse the effect of an action the smaller its impact.

(b) The case of a change in demand for good $B$ corresponds, *mutatis mutandis*, to case (a) studied above.
(c) *A change in the trajectory of the curve of costs*, i.e. a change in the form of function $F(x+y)$ produces interesting correlations between prices. A disturbance of this kind is resolved insofar as there is now for the same amount of production a corresponding cost that is different from the initial cost; in other words for the same cost there is a corresponding level of production different from the initial level.[7]

Evidently if cost falls global production will increase and this increase will be followed by a simultaneous fall in the prices of $A$ and $B$. If on the other hand cost rises global production will fall and the two prices will rise simultaneously. Hence in contradistinction to what happens in the case of a change in demand for one of the two goods, where there is a change in the global cost the prices of the two goods move in the same direction. The extent of these movements depends (with all other conditions the same) on the steepness of the new curve of costs at the points close to the new equilibrium position, and they are in fact less substantial as the curve is steeper and vice versa. This principle is so self-evident that it does not require any special demonstration.

But if we are to gain a proper understanding of the correlations between prices in the present case we have to establish both the absolute extent of movement of the two prices and the relative force of their movements; in other words we have to establish which of the two prices changes more and which less in the various cases. If all other conditions remain unchanged the relative force of the movements of the two prices depends on the elasticity of the two demands and on the quantitative relation between the production levels of the two goods, i.e. on the value of the constant $k$. If demands for the two goods have the same elasticity, and if $k=1$, i.e. $x=y$, i.e. if the two goods are produced in equal quantities, the shift in the two prices is proportional to the prices themselves. If with the elasticity of the two

demands remaining the same $x > y$, i.e. $k > 1$, then the shift in the price of $A$ will be proportionately greater than the shift in the price of $B$. If where $k = 1$, i.e. $x = y$ demand for $A$ is more elastic than demand for $B$, the shift in the price of $A$ will be proportionately less than the shift in the price of $B$. If with elasticity of demands for $A$ and $B$ different $x$ is greater than or less than $y$ the shifts in the two prices will in the great majority of cases be proportionately different. But the effects of the differing elasticities and of the differing proportions of production add up, i.e. work together cumulatively to accentuate the difference in the relative shifts in the two prices as long as the good that is subject to the most elastic demand is at the same time the good produced in a smaller quantity.[8] But the effects of the difference in elasticity and production negate each other, i.e. tend to cancel each other out or to attenuate the difference between the relative shifts in the two prices, as long as the good that is subject to the most elastic demand is at the same time the good produced in greater quantity.[9] These points leave open the possibility of a highly exceptional but theoretically possible case in which the difference in the elasticities of the two demands and the differences in the quantity produced are such and combine together in such a way that their effects balance one another out exactly; here therefore – in spite of the different elasticity of the respective demands – the shifts of the prices in the same direction are exactly proportional to the prices themselves.
(d) Let us now move on to a consideration of the effects of the *introduction of a tax*. Suppose for example that a tax on manufacturing is levied at so much for each unit produced. Let us as usual assume two goods $A$ and $B$ produced jointly and let us suppose that their production is in equilibrium. Now let us suppose that a tax of the type indicated is introduced. What are the effects of this and what are the transposition phenomena? In the specific case under examination a tax of this kind may be levied in different ways – i.e. it may be proportional

(*I*) to global production;
(*II*) to the production of $A$;
(*III*) to the production of $B$;
(*IV*) to the production of $B$ and $A$.

(*I*) In case *I* the tax resolves into an increase in the global unitary cost of production so it leads to a fall in production and a rise in the two prices with all the possible combinations considered above in respect of case (c).

(*II*) In case *II* the tax, set at so much per unit of *A*, is proportional in its overall amount to the production of the said good. To indemnify himself for the tax the producer will raise the price of *A* so as to make the consumers pay the tax, or else the rise in the price of *A* will reduce demand and hence reduce production. But at the same time production of *B* will also decrease and its price will rise. Hence a tax on the production of *A* ends up by raising the price of both the jointly produced goods.
(*III*) This case is just the same as the preceding one.
(*IV*) If instead of being applied exclusively to *A* or exclusively to *B* the tax falls on both the two products (i.e. its unitary amount weighs partly on *A* and partly on *B*) the price of both will rise at the same time, but – rather than being proportional to the unitary amount of the respective tax – the rise will be greater or smaller depending on the other conditions of the equilibrium.

This rather summary study of the general effects of a tax of the kind under discussion in the various modes in which it may be applied, suggests two important conclusions:
(a) That however the tax is introduced its general effects will be the same; it will produce a rise in the price of both goods, a rise which, assuming parity in the unitary amount of the tax, will differ according to the mode of application of the tax.
(b) That a tax on one or two or more jointly produced goods not only affects the consumers of the good that is being taxed, but also the consumers of the other good or goods that are exempt from the tax. With parity of unitary tax and with all other conditions unchanged, the rise in the price of goods exempt from the tax is smaller to the degree that the number of goods produced jointly is greater and therefore the number of exempt goods is also greater by comparison with the good or the goods that bear the tax.

But there are anomalies about these repercussive phenomena that become clearer as we probe deeper. The inquiries that follow refer to cases *II*, *III* and *IV* – there is no need for further study of case *I* because the conclusions in section (c) above apply fully to it.

As a general rule let us first of all note that where there is a tax of the kind under discussion there are two prices of the good affected by the tax that need to be taken into consideration – the price after tax and the price before it. The first is the price paid by the consumers and the second is the unitary net return actually accruing to the producers. The first influences demand and the second supply.

On this basis it seems obvious first of all that in the case of joint goods as in the case of independent products the disturbance resulting from the tax is smaller insofar as costs are more closely governed by the law of increasing costs. In fact in all three cases under consideration the tax when first introduced resolves for the producer into a reduction in the net price of the good affected, a reduction that thereby causes a corresponding fall in the *global net unitary price*. Now if before the introduction of the tax the global price coincided with the global unitary cost, after the introduction of the tax the global price is lower than the global unitary cost. The result is that production decreases, causing a rise in the two prices after tax and consequently in the global net price. Now if the global cost falls rapidly with the fall in production, i.e. if the curve of costs shows a steep rise, the global price after tax will only have to go up by a small amount for the global net price to coincide with the cost once again. For this to happen it will be enough for production to decrease by a small amount. Hence equilibrium will be reestablished by a small decrease in production and if all other conditions remain the same by a small increase in the two prices after tax.

However if costs go down slowly with the fall in production, the global price after tax will have to go up a great deal before the global net price once again coincides with the cost, and this requires a significant fall in production. Hence disturbances of prices resulting from the tax are greatest in the case of constant or slowly rising costs and smallest in the case of rapidly growing costs.

Let us now consider the relative shifts in prices of two goods in the case where just one of them is taxed; here we shall encounter some singular anomalies. I shall not attempt to cover all possible cases and combinations but will restrict myself to putting forward one extreme case in which the anomaly is the most glaring. Let $A$ be the good subjected to a tax, and $B$ the good that is exempt from it. Let demand for $A$ be very elastic and that of $B$ inelastic; let $K < 1$ i.e. let the quantity of $B$ that is obtained from the production process be much greater than the quantity of $A$. The tax is put on $A$. The net price of $A$ falls and the producer will attempt to recoup by raising the selling price of $A$. But even small price increases reduce the consumption of $A$ and its production therefore decreases. But a small decrease in the production of $A$ is accompanied by a large decrease in the production of $B$, with the result that because of the inelasticity of demand there will soon be a substantial rise in the price of $B$. Ultimately in the new equilibrium position the price of $B$ will be much higher than the initial price while the price of $A$ on the other hand will rise much less. Hence in this case there will be the apparently paradoxical result that a tax on just one of two goods produced jointly raises the prices of goods exempt from it much more than it raises the price of the good on which it is actually levied. A similar anomaly can be seen in case *IV.* Let us suppose that the tax is applied simultaneously to $A$ and $B$, but let the tax on $A$ be much higher than that on $B$. If at the same time all the other conditions we have assumed remain the same the consequence of the introduction of such a tax will be that the price of $B$ will rise more than that of $A$. Hence in this case we shall have the very weird result that where two goods are produced jointly but taxed differently the good carrying the heavier tax will rise in price less than the one carrying the lighter tax.

Another rather interesting anomaly arises in the very special case of a tax so devised that whatever way it is applied it produces the same fiscal return. For this to happen it is necessary that the unitary amount of the said tax be variable in relation to the number of units of the good to which it is applied, so that in different cases it gives an overall return that is always constant. Strange though it may seem, such a tax is indeed conceivable and it can be considered as an example of a quota tax. This can be applied either to global production, or to the production of $A$ or to the production of $B$, or to the production of $A$ and $B$ at the same time. What are the results in these various cases? With regard to cases of a fixed unitary tax we have already seen how the way in which the tax is applied is not of capital importance with respect to the disturbances it causes, for in certain cases it may happen that the good exempt from the tax rises in price more than the good to which the tax is applied; this anomaly means that the mode of application is a matter of no importance in the case under consideration.

A tax of the type under discussion yields a constant return. The equilibrium established after introduction of this tax will be of such a kind that the new global net price coincides with the new global unitary cost, taking account in the net price or in the cost (depending on the case) of the unitary amount of the tax applied. But what matters here is the overall amount of the tax – which *per hypothesi* is always at the same level in the different cases – and it is this and not the unitary tax that must be taken into account. Since it is the overall amount of the tax that counts we have to adopt a different approach to the other factors in the equilibrium; instead of tackling the problem on the basis of unitary prices and costs we shall now tackle it on the basis of total returns and costs. Before the introduction of the tax, equilibrium was established at prices of *A* and *B* such that in relation to the quantities produced the total gross return on *A* plus that of *B* corresponded exactly to the total global cost.[10] After the introduction of the tax equilibrium is disturbed. The tax will resolve into a rise in the total cost if commensurate with global production, into a reduction in the gross return on *A* or *B* if commensurate with the production of *A* or *B*, or into a reduction in the gross return on *A* and *B* if distributed between *A* and *B*. But whatever way it is imposed the end result is the same.[11] Whether it resolves into a rise in the total global cost or into a reduction in the total gross return, the equilibrium between cost and total return is disturbed in the same direction. And since the overall amount is always the same whatever way it is applied, the disturbance that results in the various cases not only takes the same direction but has the same force. If the disturbance is the same, the process of adjustment of prices and production necessary for the recovery of equilibrium is also the same in the various cases. This means that however the tax is applied the new equilibrium is established through an identical movement of prices and an identical fall in production. Now a tax of the kind being discussed is a tax which, always identical in amount in the various modes of application, exerts the same fiscal pressure in the various cases. Hence in the case of joint costs, if there is parity of fiscal pressure, the manner of application of the tax has no bearing on the resulting disturbance. In relation to prices linked through the agency of supply, this conclusion may be regarded as a particular case of the more general law proposed by Pantaleoni in respect of an economic system in which all the curves of utility and costs are interlinked; this law states that whatever the manner in which taxes are applied and whatever the goods to which they are applied, the resulting disturbance (assuming parity of fiscal pressure) is always the same.[12]

(e) From the point of view of a scientific treatment the case of a premium on production is very similar to the case of a tax, despite the fact that a premium has exactly the opposite ultimate effects. A comparison may be made on the basis: (1) of global production; (2) of the production of *A*; (3) of the production of *B*; (4) of the production of *A* and *B* together. The same disturbances and the same kinds of repercussive effects noted for taxes also occur, *mutatis mutandis*, with premiums on production. The evident analogy makes it unnecessary to provide proofs like those already given for taxes, and it will therefore be sufficient to expound the conclusions in axiomatic

form, leaving to the reader the task of checking out their accuracy. The conclusions are as follows:
(1) Whatever the basis on which the premium is allocated the general effects are the same: it encourages the expansion of global production and a fall in the price of both products.
(2) A premium applied to the production of a good produced jointly with others turns to the advantage both of consumers of the good bearing a premium and of the consumers of the other good or goods.
(3) The expansion in global production and the consequent overall fall in prices deriving from the introduction of a premium are the more substantial to the extent that global production conforms the more closely to the law of decreasing productivity.
(4) Given a premium applied to the production of a good produced jointly with another, the fall in the price of the good exempt from the premium will be more substantial than the fall in the price of the good bearing the premium whenever demand for the exempt good is less elastic than demand for the premium-bearing good, and whenever production of the exempt good is greater for each unit of global production than that of the premium-bearing good.
(5) Where the sum paid overall by the state as a premium on production is always the same, the manner in which the premium is applied has no bearing on its ultimate effects; that is to say whether it is allocated on the basis of global production or on the basis of production of one or other product the resulting disturbance is the same and the new equilibrium prices are identical in every case.
(f) Now let us consider the case of a change in constant $k$ arising from technical progress in the production process. Let us suppose that this progress is of such a nature that it results in a production of good $A$ that is quantitatively greater than before while that of good $B$ is smaller, so that constant $k$ becomes $k_1$. What will be the end result? The discovery of the new and technically superior production process does not automatically exclude the possibility of retaining the earlier production process, and the new process will be adopted only insofar as it is more profitable. This will be the case only in given conditions and within certain limits. Here let us suppose *per hypothesi* that the transformation of the production process does not cost anything: though this is an unreal hypothesis, it is legitimate on grounds of simplification in pure economics.[13] On this basis then what are the necessary conditions for the substitution to be possible and profitable? With the adoption of the new production process – which we shall now refer to for purposes of brevity as $k_1$ – the initial proportion

between $x$ and $y$ changes from what it was in the process $k$. Let us suppose that $x$ increases in proportion $m$ while $y$ decreases in proportion $n$. If in the initial production process in correspondence with a given global production $(x_0 + y_0)$

$$\frac{x_0}{y_0} = k$$

then in correspondence with the same global production[14] we shall now have:

$$\frac{x_0 + mx_0}{y_0 - ny_0} = k_1$$

The introduction of the new production process, leading to a rise in the production of $A$ and a fall in the production of $B$, produces a fall in the price of $A$ and a rise in the price of $B$ (provided there is parity of global production); that is it produces a complete rearrangement of individual prices which will probably (though not necessarily) lead to a change in the global unitary price. Now the introduction of the new production process will be profitable whenever the producers, adopting the new process as a replacement for the previous one, obtain a greater than normal profit: this occurs every time that the new global unitary price – corresponding to the initial global production – is greater than the corresponding global marginal cost. But in respect of cost two cases may occur; (1) the curve of costs relative to the new process is completely identical with the initial curve; (2) it is different. In the first case the profitability of adopting the new production process will be clear whenever the new global unitary price is greater than the initial global price or in other words than the initial global marginal cost;[15] in the second case whenever the new global unitary price is greater than the new global marginal cost corresponding to the initial extent of global production.[16] While these inequalities exist the new production process will be substituted for the old, which will lead to an expansion in global production, an expansion that will continue until – thanks to the resulting convergent movement of prices and costs – the new global unitary price coincides with the new global marginal cost.

Whether or not these inequalities arise depends on the nature and behaviour of the equilibrium factors in the two positions referred to. Hence to discover the conditions propitious for the introduction of the new process it is necessary to break down the phenomenon under investigation into these prime factors. Putting aside for the moment the possible differences in the global price, we must bear in mind that this price is made up of two distinct parts – a fraction of the price of $A$ and a fraction of the price of $B$ – and that two factors combine to determine the absolute amount of each of these two parts: the price of the respective good and the quota part of that price which forms part of the global price. Now with the introduction of the new production process the price of $A$ falls while that of $B$ rises, but at the same time, just as production of $A$ expands and that of $B$ contracts, the quota part of the price of $A$ rises and that of $B$ falls. Thus a change in a given direction of the price of either one of the two goods is met by a change in the opposite direction of the quota part of each price that enters into the formation of the global price. This prevents us from saying *a priori* in which direction each of the two parts of the global price actually changes and therefore in which direction the global price itself ultimately changes. Evidently it all depends on the amount of the rise or fall in each price in relation to the amount of the decrease or increase in the part of the price under discussion. Now the amount of the rise or fall in each price depends on the elasticity of demand for each good and on the absolute variations in the quantities of these goods produced. The amount of the decrease or increase in the quota part of the price of each good however depends on the absolute variations in these quantities.[17] And the variations in these quantities depend in turn both on the percentage increase or decrease in the production of each good, and on the initial quantity of each good on the basis of which that percentage is calculated, or on the quantity of each good produced in the first process. Hence the occurrence or otherwise of a change in the global price in a direction favourable to the replacement of the production process depends on three fundamental elements: (1) the value of $x_0$ and $y_0$ and therefore the value of $k$; (2) the value of $m$ and $n$; (3) the elasticity of demand for $A$ and $B$.

To study in what way and under what conditions these three elements favour the introduction of the new production process it will be helpful to consider the total return instead of the global unitary price – something that in every other respect amounts to the same thing. Obviously we shall need to contrast with this the global total cost rather than the global unitary cost. Staying with the hypothesis assumed up to now, we shall suppose this cost to be identical in the two production processes as long as there is parity in global production. For the new process to be introduced it is necessary that with global production the same the total return on the new production process be higher than on the previous one. The total return is the product of the unitary price and global production and like that price it is composed of two distinct parts: the partial return on *A* and the partial return on *B*. Each of these partial returns comes from the product of the price of the respective good and the respective quantity. Now in the transition from one production process to the other the price of *A* goes down but its quantity increases while the price of *B* goes up but its quantity decreases. For the total return then to be greater than the previous return it is necessary that changes in the prices and quantities be such and combine in such a way that the two partial returns increase or that the one increases and the other decreases but that the increase in the first be greater than the decrease in the second. Now the return on a good whose quantity increases while its price falls increases with the increase in the quantity when the elasticity of demand for this same good is greater than one. The return on a good whose quantity is falling but whose price is rising increases with the decrease in quantity when the elasticity of demand for the said good is less than one. Hence both partial returns will increase whenever elasticity of demand for *A* is greater than one and that of *B* is less than one, so that when these conditions are in place the total return will most certainly increase. If however elasticity of demand for *A* is less than one and that of *B* greater than one then both partial returns will fall and the total return will fall too. But any certainty as to the definitive result disappears in the intermediate combinations where elasticity of demand both for *A* and for *B* is greater than or less than one. In fact in the first of these two combinations while the return on *A* grows that on *B* decreases, and in the second case while the return on *A* decreases that on *B* increases. In these conditions it is obviously impossible to be absolutely sure whether the total return will ultimately grow or decrease, since all depends on the prevalence of an increase or decrease in the two partial returns. But if any certainty disappears there still remains the possibility of evaluating the conditions that make a change in the total return in a given direction more or less probable. In the first of the two combinations the total return increases every time the increase in the return on *A* is greater than the decrease in the return on *B*. And this is all the more likely to happen the greater the increase in the price of *A* and the smaller the decrease in the price of *B*.

Now the increase in the return on A is greater, when the elasticity of demand for *A* is greater than one and the increase in the quantity of *A* is the more substantial. The decrease in the return on *B* is the smaller, the more the elasticity of demand for $B$ is greater than one and the less substantial is the decrease in the quantity of $B$. The increase in the quantity of $A$ is the greater as the value of $x$ and $m$ is greater, and the decrease in the quantity of $B$ is so much the smaller as the value of $y$ and $n$ is lower. Hence, leaving aside any consideration of the respective elasticity of demand, a rise in the global price is the more probable, the more $x > y$, i.e. the more $k > 1$ and the more $m > n$.

In the second of the two combinations the total return increases whenever the decrease in the return on $A$ is smaller than the increase in the return on $B$, and this of course is likely to happen when the conditions are precisely the reverse of those indicated above.

These are the conclusions I have been led to by my research on the composition of and changes in total returns. But the problem of the search for conditions favourable to the introduction of a new production process as laid down at the beginning of the present study all relate practically exclusively to the global unitary price. So to provide a solution to the problem adequate to its initial form we now need to adapt the conclusions reached in respect of total return to the global price. And this does not present any difficulty, for the global price is only the total return corresponding to the unit of global production. Therefore with total global production remaining the same – a condition I have always assumed to be implicit in my study – anything that pushes the total return up or down pushes the global price up or down proportionately so that the conclusions reached above are valid for the global price too without any substantial change. If therefore we return to the global price and relate the results above to it, we shall obtain the solution of the problem at issue. This solution can be put in the four following propositions:

*I.* Whenever elasticity of demand for $A$ is greater than one and that of demand for $B$ is less than one the global price resulting from the introduction of a new production process is *always* higher than the preceding one and the introduction of the new process is *always* possible and profitable.

*II.* Whenever elasticity of demand for $A$ is less than one and that of demand for $B$ is greater than one the global price falls and the introduction of the new process is *never* possible or profitable.

*III.* Whenever elasticity of demand is greater than one both for $A$ and for $B$ the rise in the global price and therefore the introduction of the new production process will be *all the more probable* (a) the more elasticity of demand for $A$ is greater than one and the less elasticity of demand for $B$ is greater than one; (b) the more $k$ is larger than one; (c) the more $m$ is larger than $n$.

*IV.* Whenever elasticity of demand for $A$ and $B$ is less than one the rise in the global price and therefore the introduction of the new production process will be *all the more probable* (a) the less elasticity of demand for $A$ is less than one and the more elasticity of demand for $B$ is less than one; (b) the more $k$ is less than one; (c) the more $m$ is less than $n$.

These conclusions are valid for the case where the global unitary cost remains unchanged in the transition from one production process to the other – assuming the same level of production – but we have already seen how this cost can and indeed generally does in fact change. Hence to complete the study of this subject we must now also take into consideration the hypothesis of a changed cost. On this hypothesis the conditions requisite for the replacement of the old production process are modified in certain respects but do not change fundamentally; all that changes is the level of certainty or probability that the replacement will actually happen. In the transition from one production process to another the unitary cost may rise or fall. Where there is a fall, since the replacement happens only on condition that the new global price is higher than the new cost it is of course all the more probable that with all other conditions unchanged the new price will be higher than the new cost if the latter falls than if it remains unchanged. If on the other hand the cost increases, it seems obvious that it is more unlikely or perhaps less probable that the new price will be higher than the new cost. All we have then is a change in the level of certainty or probability.

Therefore in the case where the unitary cost changes, the four preceding conclusions are transformed into the following: where there is a fall in the unitary cost the first conclusion remains unchanged; the second is modified in this sense that the certainty of the impossibility of the replacement becomes only a probability; the third and the fourth do not change substantially except that with all other conditions unchanged the probability that the replacement will occur grows, and it grows all the more to the extent that the fall in cost is substantial. Where there is an increase in the unitary cost the first conclusion changes in that certainty about the replacement gives way to a degree of probability; the second conclusion remains unchanged; the third and fourth conclusion are changed only in that with all other conditions the same the probability that the replacement will happen is less than it is where costs are unchanged and is so to the extent that the increase in cost is less.

(g) The case considered above where a change in the technical process brings a change in the production ratio is a reminder of a more complex case of production at joint cost where the ratio of the goods produced jointly is not fixed but variable so that it is liable to change gradually in one direction or the other without any change occurring in the technical basis of production. A case of this kind happens wherever there is a given production process which produces a principal product and waste, and in which the quantity of the principal product can be reduced or augmented in relation to the quantity of the waste depending on how far the extraction or refinement process is taken. A similar case is that of a production process from which are obtained simultaneously goods of the same kind but of varying quality. Depending on the level to which production is pushed, the ratio of prime quality goods produced to those of inferior quality is different.[18]

Clearly, where these new conditions apply the laws of equilibrium change at least partially. We shall study these laws by focusing on the case of a disturbance arising from a change in demand for a good, on the grounds that this is the most typical type of disturbance, and I shall follow my customary practice in assuming that there are two jointly produced goods, *A* and *B*. On this basis, where there is a determinate equilibrium position and then an increase in demand for *A*, there are (unlike in the previous case) two ways by which equilibrium may be reestablished: either (as before) through an increase in global production, or through a modification in the production process involving a greater production of *A* and a smaller production of *B* (assuming parity of global production). Equilibrium will of course be reestablished in the manner that is most profitable in each case, but what are the conditions that determine this greater or lesser profitability?

As soon as demand for *A* and the price of *A* increase, the global price rises above the global unitary cost, and since this gives extra profit to the producers it stimulates them to increase production of *A* in one of the two modes indicated. But as production of *A* increases, the extra profit tends to fall in both cases, though the decrease generally takes place at a different rate in each case. Now the producers want to sustain the evanescent extra profits for as long as possible and are naturally inclined to go for a procedure that will ensure that the decrease in the extra profit takes place more slowly. Hence they will choose the one of the two ways which for each successive increase in production of A leaves the producer the greater extra profit. This will then be the most profitable way for equilibrium to be reestablished.

Now the extra profit arises from the difference between the price and the global unitary cost, and the speed with which the extra profit decreases depends in turn on the speed with which the price and the cost draw closer to each other as they move towards the new equilibrium position. Hence the problem of the conditions favourable to the profitability of one or other of the ways resolves into a study of the conditions that in the two cases exercise a decisive action on the behaviour of price and cost in response to each successive increase in production of *A*.

As to the behaviour of the global unitary cost, this can be determined more easily. In the case of an increase in global production, marginal cost rises as usual. In the second case each successive intensification of production for purposes of obtaining a larger proportion of the principal product (if global production remains the same) normally involves a supplementary cost.[19] Therefore the cost rises in this case too. Of course where the element of cost is concerned the way which involves the smaller increase in it is more suitable.

Movements of the price or of the global return (which can be more readily studied at this point) are more complex. The global return in fact consists of the partial return on *A* and the partial return on *B*. With respect to the partial return on *A* there are no variations of any kind as between the two cases: with each successive increase in the production of *A* its price falls in the same manner. Hence the return on *A* corresponding to each increase in the production of *A* represents the same amount in the two cases. However the behaviour of the partial return on *B* is different in the two cases. In the first case – an increase in global production – production of *B* increases while its price falls. In the second case – a modification to the production process – the quantity of *B* decreases but its price rises. In both cases of course the return on *B* changes (other than in very exceptional conditions). But the force and direction of this change are usually different in the two cases, depending on the elasticity of demand for *B*. If the elasticity of demand for B is greater than one the return on *B* increases in the first case but decreases in the second. If this elasticity is less than one the return on *B* decreases in the first case and increases in the second. If this elasticity equals one (and this is the highly exceptional case *I* referred to above) the return on *B* remains unchanged in both cases. Now since the partial return on *A* does not differ at all in the two cases, the possible differences in the total return depend exclusively on the possible differences in the return on *B*; and the comparative trajectories of the total return in the two cases consequently depend on the elasticity of demand for *B*.

Let us now summarise what has been said thus far and coordinate the principles separately ascertained for cost and return, and this will immediately give us the solution to the problem that concerns us. This solution I will sum up in the three propositions that follow, indicating what are the conditions requisite if equilibrium, once disturbed by an increase in demand for *A*, is to be reestablished through an increase in global production, and what are the requisite conditions for it to be reestablished through a change in the production process.

*I.* If the elasticity of demand for *B* is equal to one it does not matter where return is concerned if global production is increased or its tendency modified, for everything will depend on the behaviour of cost. If the increase in global cost corresponding to any increase in the production of *A* is equal to the supplementary cost, it will not matter which of the two ways is chosen. The first way will however be the chosen one if the increase in global cost is less than the supplementary cost while the second way will be preferred in the opposite case.

*II.* If the elasticity of demand for B is less than one it will be more advantageous in terms of return to go for a change in production than to increase global production, but the ultimate result depends on the comparative behaviour of the two costs. If the increase in the global cost is greater than the supplementary cost there will be even more pressing justification for the solution indicated. If the increase in the global cost is less than the supplementary cost a change in production policy will be preferred to the other expedient only on condition that the increased return inherent in it is greater than the higher cost it involves. In the case where the greater return is equal to the greater cost the solution will be without significance.

*III.* If the elasticity of demand for B is greater than one it will be more advantageous where return is concerned to have recourse to an increase in global production, but in this case too the ultimate result depends on the comparative behaviour of the two costs. If the supplementary cost is greater than the increase in global cost the solution indicated will impose itself with greater justification. If the supplementary cost is less than the increase in global cost it will be more profitable to increase global production than to adopt the other expedient only in the case where the better return it involves will be higher than the greater cost.

In the case where the higher return is equal to the increased cost, it does not matter which of the two expedients is chosen.

So it emerges from the study of a case of joint production in which the ratio of the jointly produced goods is liable to be modified that an increase in demand for one of the said goods causes either an increase in global production or a change in the ratio depending on whether conditions favourable to one or the other solution are prevalent. But whereas with small shifts in demand we may suppose that equilibrium is reestablished by means of just one of the two expedients, this cannot be the case where the shifts are substantial. In fact the possible change in the ratio is self-limiting and the increase in the production of *A* that may derive from it cannot be substantial enough, so even if we admit that initially the expedient in question is adequate to meet an increase in demand for it, it is clear that of itself it is insufficient where the increase is a big one. After having produced all the results it is capable of it will have to be followed by an increase in global production. But even independently of these considerations, valid only where the first way adopted is that of a change in the ratio, other general considerations point us to the same conclusions. Substantial shifts in demand for *A* imply that the new equilibrium is established at a point on the curve of demand for *B* and on the curve of costs that is a long way from the initial equilibrium points. Now the elasticity of the curve of demand for a good changes, often significantly, between one point and another on the curve, especially if the two points are distant from one another. Moreover the curves of global cost and supplementary cost do not have the same trajectory throughout but follow independent courses so that now one and now the other shows the most steeply rising increase. The result is then that after a significant increase in demand for *A* the process of recovery of equilibrium initially favours one route but along the way changes to the other (though later it may return to the first) as the comparative conditions of costs or the elasticity of demand for *B* undergo changes. In other words the two expedients, which act separately and in turn depending on the case when there are minor shifts in demand, act together and combine when there are substantial shifts in demand, so that the resulting new equilibrium is established through their joint operation.

But because in this case equilibrium is reestablished through the combination of both expedients the laws that govern it are rather different from those that regulate equilibrium in the case of a constant ratio. First of all it is evident that since the expedients combine for the recovery of equilibrium, this is reestablished (if all other conditions are the same) through an increase in global production that is smaller than in the other case. Now a smaller increase in global production implies a smaller increase in production of *B*, but at the same time the alteration of the ratio in favour of *A* contributes in another way to weakening the effective increase in *B*. Hence for two reasons the increase in production of *B* in the case of a variable ratio is less than in the case of a fixed ratio. Now since the depression in the price of *B* (if all other conditions remain unchanged) is the greater as the increase in its quantity is the greater, it is clearly greater in the case of a fixed ratio than it is in that of a variable ratio. It follows then that with a given increase in demand for *A*, in the case of a variable ratio the new equilibrium is established through a shift in global production and in the price of *B* that is smaller than in the case of a fixed ratio. The variability of the ratio thus exercises a kind of compensatory action, lessening the extent of the disturbance and making the prices of *B* less sensitive to movements in the prices of *A*. And this is natural. The case of a variable ratio is intermediate between the case of a fixed ratio and the case of independent production where the ratio is non-existent. In this latter case the prices of the two goods are independent, at least where production is concerned.

Hence it is natural that the intermediate case reflects this intermediate level of interdependence that lies between the greatest interdependence existing in the case of a fixed ratio and absolute independence as in the case of a non-existent ratio.

## II UNDER A MONOPOLY REGIME

**1.** Let us now proceed to study the case of a monopoly. The principles that govern prices of goods under a monopoly are too well known for it to be necessary to refer to their fundamentals. It will be enough here simply to develop these principles and apply them to the special case. For simplicity's sake we shall suppose as before that there are just two joint goods. The monopoly conditions in which the two goods are jointly produced give rise to two distinct cases that need to be studied separately:

(1) a monopoly of production and sale of both goods;
(2) a monopoly of production and sale of one of the goods and of competition between the two producers in the sale of the other.

This second case presupposes the existence of markets that are linked with respect to the good that is subjected to competition. Here then we have a case that falls outside the conditions of a closed market that we assumed as a basis for these initial researches, but this case will be studied later when we deal with partially or completely linked markets. Discussion in this chapter will refer exclusively to the first case.

**2.** Every perfect monopolist tends to maximise his own net profit, and this general principle is valid for the specific case of production at joint cost. The monopolist's profit comes from the difference between the total gross return and the total cost. The total gross return in the case of two goods produced jointly comes from the sum of partial returns on $A$ and $B$. It is precisely this difference that the monopolist tends to maximise. Hence if as usual we use $f(x)$ for the price of $A$, $\phi(y)$ for the price of $B$, and $g(x+y)$ for the unitary cost,[20] the problem of the equilibrium of production and prices comes down to determining the values of $x$ and $y$ that maximise the formula:

$$xf(x) + y\phi(y) - (x+y)g(x+y)$$

This makes it absolutely clear that since the two lots of partial gross returns and the total cost together constitute the maximum profit, the way in which equilibrium is established depends on the trajectory of the curves of partial returns on $A$ and $B$ and on the curve of total costs. These curves derive respectively from the curves of demand for $A$ and $B$ and from the curve of unitary costs, and it is easy to explain how they are constructed.[21]

The functions $f(x)$, $\phi(y)$, and $g(x+y)$ represent the coordinates of the curves of unitary prices and costs expressed as a function of the respective quantities. By multiplying these functions by the respective variables we obtain the functions proper to the expression under discussion, with successive values representing respectively the partial gross return on $A$, that on $B$, and the total cost, corresponding to quantities $x$, $y$ and $x+y$. Hence with their extremities the coordinates corresponding to these successive values trace out three curves – the curve of the return on $A$, the curve of the return on $B$, and the curve of the total global cost. We know the trajectory of the curves of the two returns – they are called *sale curves* and almost invariably an initial ascent is followed by a descent from a point of maximum elevation,[22] though the precise trajectory of each curve depends on the elasticity of demand for the goods in question.

But the curve of total costs calls for further explanation. As we have seen total cost comes from the product of unitary cost and global production. Now depending on the particular point reached in the company's development, unitary costs may be growing, constant or decreasing,[23] and the trajectory of the curve of unitary costs naturally influences the trajectory of the curve of total costs. If unitary costs are growing or constant then total costs are growing in each case. If unitary costs are falling the total cost will be growing if the curve of unitary costs is less steep than an equilateral hyperbole; it will be constant if this curve has the same trajectory as an equilateral hyperbole; it will be falling if the curve is steeper. But we need to notice immediately that these two latter cases are actually of secondary importance. For they assume a trajectory of the curve of unitary costs that is rapidly falling, something that presumably never happens either in theory or in practice, at least not for any substantial period of time. Hence as a general rule the normal trajectory of the curve of total costs is upwards throughout its course.[24]

**3.** The position at which the monopolist obtains maximum profit depends on the behaviour of the curve of total costs and the curve of partial returns. As we have seen total profit comes from the overall return *minus* the total cost, and the overall return comes in turn from the sum of the individual partial returns. Hence if to the coordinates of the curve of return on *A* we add the corresponding coordinates of the curve of return on *B*, we obtain a third curve, that of total returns, and from the point of view of its constitutive elements this curve will now become the basis of this study.

Once we have put the problem in this way its general solution is easy to show. For it is obvious that within the limits where total return is higher than cost it is profitable for the monopolist to increase production and sales whenever and as long as the increase in the total return for each increase in production is greater than the corresponding increase in cost, i.e. whenever and as long as the inclination of the curve of total returns is steeper than that of the curve of costs. This ceases to be profitable however as soon as the inclination of the two curves coincides either because of the inflexion of the curve of returns or because of the greater steepness of the curve of costs. The point at which this applies is the point corresponding to conditions of maximum net monopoly profit, and this then gives at least a part of the answer to our problem.

But this simple general solution becomes somewhat complicated in particular cases, and it is from the specific characteristics of production at joint cost that these complications arise.[25] In fact the monopolist of jointly produced goods finds himself tied to two conditions: to begin with he cannot produce the two goods other than in a determinate ratio, and in the second place he must regulate the production and sale of these goods so as to obtain the greatest net profit. Now as will shortly be seen these two conditions are not always compatible, or rather they become compatible in some cases only on the supposition of other conditions that differ from case to case. Therefore the problem requires different solutions to each case and these need to be examined. My research leads me to the conclusion that the following principles apply: (a) that the monopolist cannot sell his own products other than in a *closed market* in which he exercises his monopoly; (b) that he cannot or will not adopt the system of *multiple prices* and must therefore retain a *single monopoly price* for each product.

In order to resolve our problem on this basis we must first of all seek the principles that regulate the total gross return, and this will enable us easily to deduce the laws of production and price equilibrium.

As we have seen, the total return is the sum of the partial returns, and it will grow as long as they grow. The point of maximum total return depends on the two points of maximum partial return.

The maximum partial return on $A$ is obtained in correspondence with the sale of a certain quantity of $A$ that we shall call $x_0$; the maximum partial return on $B$ in correspondence with the sale of a certain quantity of $B$ which we shall call $y_0$. Now these two quantities are independent of one another, since at least in our hypothesis the curves of demand for the two goods and therefore the curves of partial returns derived from them are independent of each other. But this is not true of the quantities $x$ and $y$ of $A$ and $B$ that are produced jointly, which stand to each other in a fixed relation $k$. Now this contrast between the independence of the quantities corresponding to the points of maximum partial return and the reciprocal dependence of the quantities produced is responsible for the complications to which I have referred, complications that are reflected in the formation of the total return.

Let the formation of the total return in its ascendent phase be assumed to follow in correspondence with successively growing increases in production. It is evidently in the interest of the monopolist that the total return corresponding to any increase in production be as high as possible, and its formation conforms to this principle.

This total return is as we have seen the sum of the partial returns, and for this total gross return to be the highest possible it is necessary for the individual partial returns that make it up also to be as high as possible. But they vary in relation to the quantities of the two goods *A* and *B* put on sale, and so the quantities of *A* and *B* put on sale in correspondence with each succeeding increase in production must be such as to provide the highest possible partial returns. Therefore our study of the principles regulating the formation of the total return reduces to a study of the way in which the quantities of *A* and *B* put on sale combine together with each succeeding increase in production.

As global production gradually grows production of *A* and production of *B* increase in parallel. Now at the beginning and while the production of the two goods is still limited each succeeding increase in the production and sale of *A* and *B* is matched by increases in their respective returns. Therefore initially the entire quantity of both goods produced must be put on sale in order to obtain the highest possible total gross return. In the first part of its ascendent phase therefore the total return consists in every case in the sum of partial returns obtained by putting on sale the entire quantities of *A* and *B* successively produced.

But sooner or later, if expansion of global production continues, the moment comes when the production of *A* or the production of *B* attains to the dimensions that correspond to its own maximum partial return; i.e. to the dimensions of $x_0$ or $y_0$ respectively. For the moment we shall suppose that this happens for good *A*. Since the production of *A* cannot increase unless production of *B* increases at the same time, and since production of the two goods proceeds on the basis of a constant ratio, at the moment when production of *A* reaches quantity $x_0$, production of *B* will reach some quantity $y_1$. Now in relation to this quantity $y_1$ three different things can happen, giving rise to three different cases. It may happen that $y_1 = y_0$, i.e. that the quantity of *B* produced coincides exactly with the quantity of maximum partial return on *B*. In this case the maximum partial return is reached at the same time for both goods, and the maximum total return is obtained in correspondence with the two maximum partial returns. In this case therefore throughout its ascendent phase the total return is constituted by the sum of the partial returns obtained by putting on sale the entire quantities of *A* and *B* produced successively.

It may however happen that at the moment when the production of *A* reaches quantity $x_0$, the quantity $y_1$ of *B* that corresponds to it is less than $y_0$. In this case the point of maximum partial return is reached by good $A$ before it is reached by good $B$ and complications immediately arise. In fact once the point of maximum return on $A$ is reached, further increases in sales of $A$ cease to be profitable and are suspended. But since the quantity of $B$ produced concomitantly is lower than the quantity corresponding to its maximum partial return, further increases in sales of $B$ push up the total gross return, and the point of maximum total return is beyond the point of maximum partial return on $A$. But in order to able to sell further quantities of $B$ it is necessary to produce them, yet when $B$ is produced $A$ is produced at the same time, and since these further quantities of $A$ cannot be put on sale they must be destroyed. Hence beyond the point of maximum partial return on $A$ the total return is formed purely by means of the contribution made by the partial return on $B$, and the point of maximum total return coincides with the point of maximum partial return on this good.

Finally it can happen – and this is the last of the possible cases – that the point of maximum partial return is reached by $B$ before it is reached by $A$. In this case the same thing happens for $B$ that we have just seen happening for $A$, and in this case therefore beyond the point of maximum partial return on $B$ the total return is formed purely by means of the contribution made by the partial return on $A$, and the point of maximum total return coincides with the point of maximum partial return on this good.

On this basis we shall now be in a position to give a general form to our conclusions. In the first case the points of maximum partial return are reached at the same time for $A$ and for $B$. This happens because the quantities $x_0$ and $y_0$ of $A$ and $B$ corresponding to their respective maximum partial returns are in precisely the same ratio $k$ to one another as the quantities of $A$ and $B$ that are successively produced. Hence if we use $h$ for this ratio $\frac{x_0}{y_0}$, we see that the points of maximum partial return are reached at the same time when $h = k$.

In the second case the point of maximum return on $A$ is reached before that on $B$, and this happens because $y_1 < y_0$. Now if $y_1 < y_0$ this means that to obtain the maximum return on $B$ there needs to be a quantity of $B$ greater than what is produced in correspondence with quantity $x_0$ of $A$; this means therefore that $\frac{x_0}{y_1} > \frac{x_0}{y_0}$ or in other words that $k > h$. Hence the maximum partial return is reached for $A$ before it is reached for $B$ whenever $k > h$.

In the third case the point of maximum return on $B$ is reached before the point of maximum return on $A$, and this happens because

$y_0 < y_1$. Now if $y_0 < y_1$, this means that to obtain the maximum return on $B$ there needs to be a quantity of $B$ smaller than what is produced in correspondence with quantity $x_0$ of the greatest return on $A$; this means that $\frac{x_0}{y_1} < \frac{x_0}{y_0}$ or in other words that $k < h$. Hence the maximum partial return is reached for $B$ before it is reached for $A$ whenever $k < h$.

From these premises we can easily deduce the general law of the total return, and this can be summarised in the three following propositions:

I. Whenever $k = h$ the total return comes throughout its ascendent phase from the sum of the partial returns obtained by putting on sale the entire quantities of $A$ and $B$ successively produced. The incremental ratio of the total return, i.e. the angle of the curve that represents it, is at every point equal to the sum of the incremental ratios or the angle of the curves of the two partial returns.[26] And the point of maximum total return coincides with the two points of maximum partial return on $A$ and $B$.

*II.* Whenever $h < k$ the total return is initially formed by the combined operation of the two partial returns, and this applies up to the point corresponding to the maximum partial return on $A$. Beyond this point the total return increases solely through the contribution of the increases in the partial return on $B$. From this point therefore the incremental ratio of the total return becomes equal to the incremental ratio of the return on $B$, the point of maximum total return coincides solely with the point of maximum return on $B$, and this total maximum return is reached by putting on sale the entire quantity of $B$ produced and destroying the quantity of $A$ produced that exceeds the quantity of maximum partial return on this good.

*III.* Whenever $h > k$ the total return is initially constituted by the two partial returns, and this up to the point corresponding to the maximum partial return on $B$. Beyond this point the total return increases solely through the contribution of the increases in the partial return on $A$. From this point therefore the incremental ratio of the total return becomes equal to the incremental ratio of the return on $A$, and the point of maximum total return is reached by putting on sale the entire quantity of $A$ produced and destroying the quantity of $B$ produced that exceeds the quantity of maximum partial return on this good.

**4.** Having established the general law of the formation of the total gross return we can easily formulate the law of production and price equilibrium in the conditions under discussion. Equilibrium is in fact established in correspondence with an amount of global production to which corresponds the maximum total net profit as a function of the total gross return and the total cost involved in it. Now this happens at the point where the incremental ratio of the total return is equal to the incremental ratio of the total cost, and since the latter is growing the point is reached before the total return reaches its highest level – i.e. as a general rule equilibrium is reached in correspondence with a global production lower than that which yields the maximum total gross return. But the total return as we have seen comes from the combination of the partial returns on $A$ and $B$, a combination that operates on the basis of the laws set out above. Hence the partial returns on $A$ and $B$ combine together in the different cases in different modes and proportions to form the total return corresponding to the equilibrium point. In the different cases therefore equilibrium is established correspondingly with a different extent of production and sale of $A$ and $B$ and according to rigorous laws that are the corollary of the conclusions in the last paragraph. These laws may be summarised in the following propositions:

*I.* Whenever $h = k$ equilibrium is established at a point to which corresponds a production of $A$ and $B$ lower for both goods than that of the maximum respective partial return, and it is established without there being any unsold residue of $A$ or $B$.

*II.* Whenever $h > k$ equilibrium is established at a point to which corresponds a production of $A$ lower than that of the maximum partial return and a production of $B$ that is more or less extensive depending on the case. And in fact:

(a) if in correspondence with the point at which the incremental ratio of the total return – i.e. the partial return on $A$ – is equal to the incremental ratio of total cost, production of $B$ is equal to the quantity of maximum partial return on $B$, production of $B$ will be equal to this quantity in the equilibrium position; in this case equilibrium will be established without any unsold residue either of $A$ or of $B$;

(b) if in correspondence with a level of global production corresponding to a production of $B$ equal to the quantity of maximum partial return, the incremental ratio of the return on $A$ is greater than the incremental ratio of the total cost, global production will expand beyond this level; in this case equilibrium will be established without any unsold residue of $A$ remaining but with an unsold residue of $B$;[27]

(c) if in relation to a global production to which corresponds a production of $B$ equal to the quantity of the maximum partial return, the incremental ratio of the return on $A$ is less than the incremental

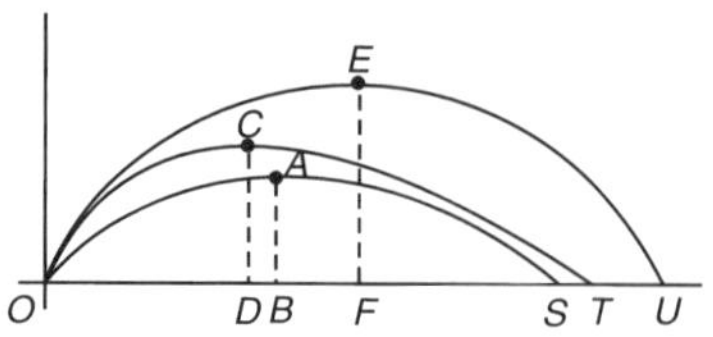


Figure 1

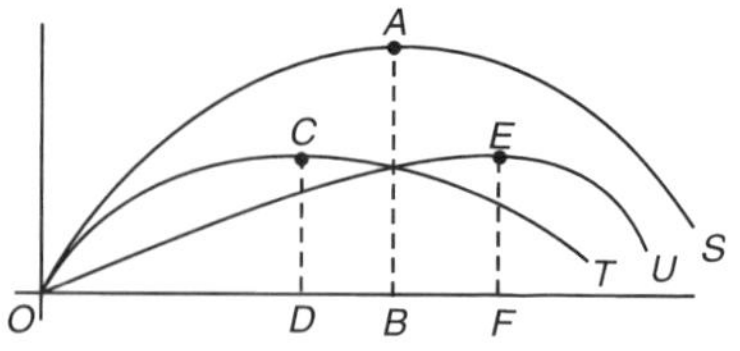


Figure 2

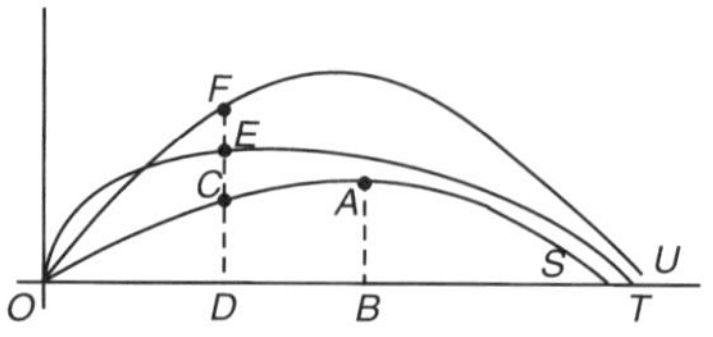


Figure 3

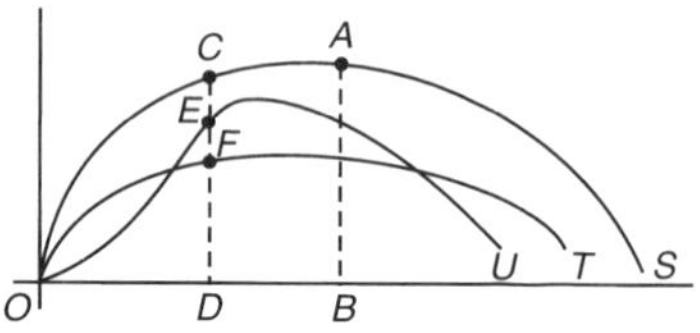


Figure 4

ratio of the total cost, global production will stop expanding before this point is reached. Equilibrium will be established without there being any unsold residue of *A* and *B* and thus with a level of production of both goods lower than that of the maximum partial respective return.

III. Whenever $h < k$ equilibrium is established at a point to which corresponds a production of *B* lower than that of the maximum partial return on *B*. Therefore in this case there will never be any residue of *B*; and where good *A* is concerned there will be a production that is equal to, greater or less than that of maximum partial return, i.e. there will or will not be a residue depending on whether the conditions set out for good *B* in cases (a), (b) and (c) apply for good *A*.[28]

In all these cases and subcases equilibrium is always exactly determined.[29]

**5.** Having ascertained and analysed the general laws of equilibrium we shall now proceed to consider how this equilibrium is recovered once it has been disturbed. This will help to clarify the laws under discussion and especially to indicate the correlations between prices of joint goods under a monopoly regime. The disturbances we shall suppose as occurring are the following: (1) a change in demand for *A*; (2) the introduction of a tax.

(1) *A change in demand for good A* results in a distortion of the curve of partial returns on *A*, a distortion that causes numerous complications. A change in demand for a good signifies a change in the quantity required correspondingly with the various prices or, rather, if we attend to the way in which our demand curves are constructed, a change in the demand prices that correspond to the

various quantities.[30] Naturally if demand grows demand prices are higher than they were at first, while if demand decreases they are lower, but these changes in the various prices can happen in different ways. The new demand prices may all be proportionately higher or lower than the initial ones, and in this case the new demand curves are parallel to the initial one, i.e. they are precisely the same initial curves but shifted with respect to the axis of the abscissae. On the other hand the new demand prices may follow a trajectory completely different from the previous ones, and in this case the new demand curve is different from the one before. This difference in the form of the new demand curve had only a secondary importance in a free market but it is of crucial importance under a monopoly regime, for on it depends the form of the new curve of the partial returns on $A$, and hence the manner in which equilibrium is recovered.

The curve of partial returns is in fact only a derivation of the demand curve, obtained by multiplying the coordinates of this – representing prices – by the corresponding quantities. Now it is evident that the trajectory of the curve of returns differs according to the trajectory of the demand curve. If the demand price goes down only a little in correspondence with the initial increases in quantity while it goes down rapidly for subsequent quantities, the curve of returns will initially rise rapidly and the point of maximum return will be reached in correspondence with a limited quantity. If on the other hand the demand price falls at a certain speed with the initial increases in quantity, while subsequently declining less steeply, the curve of returns will have a shallow rising trajectory and will reach the point of maximum return in correspondence with a substantial quantity. Now of course these two cases may occur either where prices are high or where they are low and either after a fall or after a rise in prices. Therefore whether we are dealing with a rise or with a fall in demand for $A$ either of these solutions may be found – i.e. in correspondence with a greater quantity the new point of maximum return deriving from the distortion of the first curve of returns on $A$ may be less than, or even equal to, the preceding quantity.[31]

Now the quantity of $A$ that corresponds to the maximum return is as we know one of the elements that combine to form the ratio $h$. Therefore following the change in demand for $A$, the first ratio $h$ gives way to a new ratio $h_1$. And whether demand increases or decreases this may equally well be less or greater than the first one. Hence the first tangible result of the change in demand for $A$ is potentially to change the ratio between the two quantities corresponding to the two maximum partial returns on *A* and *B*.

But if the change in this relation has a decisive influence on the new equilibrium position in the cases in which the initial position is established through a residue of *A*, in cases where this initial position is established in correspondence with a quantity of *A* produced and sold that is lower than that of the highest partial return, the change in the ratio under discussion has no influence at all. In all these cases what is really influential is the angle of the new curve of returns on *A* or in other words the incremental ratio of the new series of returns on *A* in correspondence with the quantity of *A* produced and sold in the initial equilibrium position.

In fact in any equilibrium position and therefore in the initial position that we must now consider the incremental ratio (i.e. the angle of the curve) of total returns is equal to the incremental ratio (i.e. the angle of the curve) of total costs. Now in the cases we are currently considering, the incremental ratio of the total returns in the initial equilibrium position coincides with the incremental ratio of the partial returns on *A*. Therefore if the incremental ratio of the partial return on *A* changes as a consequence of the effect of the change in demand for *A*, the incremental ratio of the total return changes. This is the source of shifts in the initial equilibrium position.

If in correspondence with the initial equilibrium point the new curve of returns on *A* is steeper than the first curve, the steepness or incremental ratio of the new curve of total returns will be greater than that of the first, i.e. it will be greater than the steepness of the curve of costs and there will be an incentive to increase production. If on the other hand, still in correspondence with the initial equilibrium quantity of *A*, the new curve is less steep than the first, the new curve of total returns will be less steep than the first one, i.e. it will be less steep than the curve of costs and there will be an incentive to reduce production. Hence in all these cases the nature of the disturbances that derive from the change in demand for *A* depends exclusively on the steepness of the new curve of returns on *A* in comparison with the preceding curve at the point indicated. The whole problem of the impact of the disturbance under discussion therefore reduces to a study of the way that equilibrium is restored after a change in the steepness of the curve of returns on A at the point corresponding to the previous equilibrium.[32]

A change in the steepness of the curve of returns at the point indicated may happen either in the case of an increase or in the case of a decrease in demand. Hence under a monopoly regime this strange anomaly may arise, by which an increase in price caused by an increase in demand is accompanied by a decrease in production; and vice versa a reduction in price caused by a reduction in demand may be accompanied by a hike in production.

On this basis we are finally in a position to formulate concrete conclusions on the way in which equilibrium, disturbed by a change in demand for $A$, is restored in the various cases. And this we shall do by making further use of the schema of cases set out a short while ago in expounding the general laws of equilibrium. The results obtained in the different cases are as follows:

*I.* We know that whenever $h = k$ equilibrium is established with a production of $A$ and $B$ that is lower for both goods than that of the maximum partial return. Now let us suppose that demand for $A$ increases. The results are different depending on the case. If the new curve of returns on $A$ in correspondence with the quantity of $A$ previously produced is steeper than the previous curve there will be an incentive to increase production. Hence in the face of the rise in the price of $A$, that of $B$ falls. The initial ratio $h$ is replaced by a new relation $h_1$. If $h_1 < 0 = k$ equilibrium will be established without any residue of $B$ remaining. If $h_1 > k$ it may happen – though not necessarily so – that production of $A$ is pushed to such a point that the concomitant production of $B$ is greater than the quantity of maximum return on $B$. In this case the increase in the price of $A$ is matched by a fall in the price of $B$ but only for a while, up to the price level corresponding to the maximum return. Beyond this level the price of $B$ remains unchanged whether or not there is any further increase in the quantity.

If the new curve of returns on $A$ is steeper at the point indicated than the preceding curve, this will be an incentive to reduce global production. Unlike in the previous case, the increase in the price of $A$ will in this case be matched by a corresponding rise in the price of $B$ and so the two prices will be linked by a direct correlative relation.

Let us now suppose that demand for $A$ falls. In this as in the previous case if the new curve of the return on $A$ is steeper at the point referred to several times above than the initial curve, production will tend to increase. In this case therefore the fall in the price of $A$ will be matched by a corresponding fall in the price of $B$. There will be no residue of $B$ if $h_1 < 0 = h$ and therefore $< 0 = k$. There may be a residue of $B$ only where $h_1 > h$ and therefore $> k$.

If however the new curve of the return on $A$ is less steep than the previous curve, there will (as in the similar case of growth in demand) be a fall in the production of, and a rise in the price of $A$.

*II.* Whenever $h > k$ equilibrium is established as we know in correspondence with a production of $A$ lower than the quantity of maximum return on $A$, and in correspondence with a production of $B$ that differs depending on the case:

(a) Production of $B$ may correspond exactly to the quantity of maximum return on $B$. Let us then suppose that demand for $A$ increases. If the new curve of returns on $A$ is steeper at the usual point than the previous one, global production will grow and there will be a residue of $B$, the price of $A$ will go up and that of $B$ will remain unchanged. If however the new curve is less steep than the previous one production will fall and the rise in the price of $A$ will be matched by the rise in that of $B$.

If demand for $A$ falls similar results may follow to those just indicated; with this difference that in the case of a rise in production the fall in the price of $A$ is matched by an unchanged price for $B$, while where there is a fall in production the fall in the price of $A$ is matched by a rise in that of $B$.

(b) Production of $B$ may be greater than the quantity of maximum return on $B$ and therefore a residue remains. Let us suppose that under these conditions demand for $A$ grows. If the curve of returns on $A$ is steeper than the previous one production will grow while the residue of $B$ will grow proportionately and the price of $A$ will rise while that of $B$ will remain unchanged.

If the new curve of returns on $A$ is less steep than the previous one global production will fall, and the rise in the price of $A$ will or will not be matched by a rise in the price of $B$, depending on whether the resulting fall in production of $B$ is greater than, equal to, or less than the initial residue of $B$.

Similar results, though with the differences indicated above, happen in the case of a fall in demand for $A$.

(c) Production of $B$ may be lower than the quantity of maximum return on B. In this case we are in many respects back with case *I* where initially $h = k$, and the reader is referred to that case for the solutions provided by the various combinations.

*III.* Whenever $h < k$ equilibrium is established with a production of $B$ lower than that of maximum return on $B$ and with a production of $A$ that depending on the case may be (a) equal to, (b) greater than, (c) less than the quantity of maximum return on $A$. In each of these cases the conditions accompanying equilibrium are as we know different. Now since a change in demand for *A* may shift the point of maximum return on *A* and in doing so may change the initial equilibrium conditions by transforming the conditions of case (a) into those of case (b) or of case (c) or vice versa, the manner in which equilibrium – once disturbed by the change in demand for *A* – is recovered is decisively affected by the resulting shifts in the point of maximum return on *A*. These shifts we must now consider in the case under examination.

The point of maximum return on *A* may shift in the direction of corresponding to a quantity of *A* greater or smaller than the previous quantity, or it may remain unchanged, i.e. correspond to a quantity of *A* equal to the previous quantity. Now the quantity of *A* corresponding to the maximum return is as we know an element in ratio *h*, which is therefore directly proportional to it. Hence depending whether after a change in demand for *A* this quantity grows, decreases or remains the same, the value of *h* will grow, decrease or remain unchanged. Therefore the study of the effects of change in demand for *A* in the various combinations currently under discussion comes down to the study of the disturbance of equilibrium that results from a change in *h* in the various combinations.

On this basis let us now proceed to study the problem in the various possible cases.

(a) In this case let us suppose that in the initial equilibrium position production of *A* is equal to the quantity of maximum returns so that equilibrium is established without there being any residue of *A*. Let us now suppose that there is a rise in demand for *A*, naturally causing a rise in its price. If *h* remains unchanged there will be no change in global production or in the price of *B*. If *h* increases global production will grow and the rise in the price of *A* will be matched by a fall in the price of *B*. But if *h*, which in the beginning is smaller than *k*, increases so as to become greater than *k*, in that case it may happen – though this is not strictly speaking inevitable – that there is a residue of *B* and that therefore the price of *B* does not fall in relation to the whole resulting rise in production of *B*. If *h* decreases however global production remains unchanged, but since the quantity corresponding to the maximum return on *A* has fallen, there will be a residue of *A*. There will be no change in the price of *B* to correspond with the rise in the price of *A*.

Similar results are to be observed in the case of a fall in demand for *A*, only in the different subcases the prices of *A* are always lower instead of higher than the previous ones.

(b) Let us suppose that in the initial equillibrium position production of *A* is greater than the quantity of maximum return and is therefore accompanied by a residue of *A*. If there is then an increase in demand for *A* the price of *A* will definitely rise, but subsequent effects will differ from case to case. If *h* remains unchanged, the other equilibrium conditions remain unchanged. If the value of *h* rises the residue of *A* decreases but there will be no increase in global production and therefore in the prices of *A* and *B* as long as the positive difference between the new quantity of *A* corresponding to the maximum return on *A* and the initial quantity is smaller than the initial residue of *A*. In this case the rise in the price of *A* will not be matched by any rise or fall in the price of *B*. As long as the above difference is greater than the initial residue of *A*, global production will rise and the price of *B* will fall. In the extreme case then in which *h* grows to become greater than *k* global production will rise, and in certain conditions it will increase to such a point as to invert the initial position and create a residue of *B*.

Finally if *h* decreases the residue of *A* will grow, but global production will remain unchanged and the price of *B* will remain unchanged.

Similar results occur in the case of a fall in demand for A – with this difference that in all possible subcases the price of *A* is always lower than the previous price.

(c) In the case where equilibrium is initially established in correspondence with a quantity smaller than that of the maximum partial return both on *B* and *A*, conditions are the same as in case *I*. And for the various solutions that are found to the problem in the different subcases the reader is referred to the discussion of case *I*.

What emerges from all this lengthy, laborious and difficult discussion is this: unlike in the case of a free market, there is no consistently strict and rigorous correlation between the prices of two jointly produced goods under a monopoly regime. Sometimes a shift in the price of *A* is matched by a shift in the price of *B*, sometimes not at all. Sometimes the price of *B* moves in the same direction as the price of *A*, sometimes in the opposite direction. Sometimes the price of *B* moves at the same time as the price of *A*, and sometimes it begins to shift at a certain point in the course of *A*'s shift, or up to a certain point in *A*'s shift. Thus correlations between the two prices are intermittent, unstable, and difficult to reduce to a general law; they are *sui generis* correlations governed by particular laws proper to each individual case and subcase. But for all that they are intermittent and unstable these correlations do most certainly exist; that can no longer be doubted after the present study. The acknowledgement of their existence is of itself enough to demonstrate the mistake made by perfectly able economists whenever they have dealt too superficially with such complex problems, providing the sort of general solutions that involve a mutual independence of production and prices of joint goods under a monopoly regime.[33]

(2) Let us now proceed to consider the effects of the *introduction of a tax*. Whenever we speak of a tax as an element that disturbs equilibrium under a monopoly regime we must of course restrict ourselves to those kinds of tax that do actually disturb a monopolistic equilibrium. It is now a matter of scientific fact that a tax which is constant in amount or proportional to the net return of the monopolist does not in any way shift the initial equilibrium position, while a shift in this position does really occur after the introduction of *a tax proportional to quantity*.[34] In this paragraph therefore I shall be referring to a tax of the latter kind, and I shall consider what transpositions and repercussions it causes – within the limits where it is proper to speak of a genuine transposition of a tax under a monopoly regime.[35]

A tax of the type indicated may in the present case of joint costs be introduced in the following four different ways:

(*I*) it is made proportional to global production;
(*II*) it is made proportional to the quantity of *A* sold;
(*III*) it is made proportional to the quantity of *B* sold;
(*IV*) it is made proportional to the quantity of *A* and *B* sold.

*I. In the case where the tax is proportional to global production*, it resolves into an increase in the unitary cost, and therefore into an increase in the total cost proportional to the quantity, i.e. an increase in the steepness of the curve of costs. Hence as in a free market the first effect of a tax here is to reduce production. But the impact on prices is different depending on the equilibrium position prior to its introduction.

(a) If equilibrium is initially established with a production of *A* and *B* lower than the quantity of maximum return, the tax, reducing production and sale of both products, will drive the prices of both up.

(b) If equilibrium is initially established with a production of *A* lower than the quantity of maximum return and a production of *B* exactly equal to it, or a production of *B* lower than, and of *A* exactly equal to, the quantity of maximum respective return, the tax will as in case (a) produce a rise in the price of both products.

(c) If equilibrium is initially established with a residue of *B* and a quantity of *A* lower than that of maximum return, as production falls the tax will first of all drive the price of $A$ upwards. The price of $B$ will remain unchanged if the new equilibrium is established with a reduction in global production involving a decrease in production of $B$ equal to or smaller than the residue; but it will rise if the tax causes a decrease in global production involving a reduction in production of $B$ greater than the initial residue. At all events however the rise in the price of $B$ will occur within the limits of the amount by which the decrease in production of $B$ exceeds the residue.

(d) If equilibrium is initially established with a residue of $A$ and a quantity of $B$ lower than that of maximum return, what will happen to $A$ is what I have shown above for $B$ and vice versa.

*II. A tax proportional to the quantity of A sold.* A tax of this kind, increasing with the increase in the quantity of $A$ sold, lessens the increase net of tax of partial returns on $A$ for each successive hike in sales. Hence the curve of returns on $A$, net of tax, is in its ascendent phase less steep than initially and the point of maximum return net of tax on $A$ is reached in correspondence with a smaller quantity than at first. Therefore the first result of the introduction of the tax is a decrease in ratio $h$, and from various points of view the final effects on production and prices may be considered entirely analogous to the effects of the decrease in this ratio.

Hence:

(a) If to begin with $h = k$ and equilibrium was established in correspondence with quantities of $A$ and $B$ smaller than those of the maximum return, the tax decreases the steepness of the curve of returns on $A$ and causes a reduction in global production, consequently making the prices of $A$ and $B$ rise simultaneously. Similar results occur if with the conditions of $A$ remaining the same, the initial equilibrium quantity of $B$ is exactly equal to the quantity of maximum return instead of being lower.

(b) If equilibrium is initially established with a residue of $B$ but with a quantity of $A$ lower than that of maximum return, the introduction of the tax will cause a decrease in global production and a rise in the price of $A$. The price of $B$ will remain unchanged if the decrease in global production involves a decrease in production of $B$ that is smaller than or equal to the initial residue; it will increase if the decrease in production of $B$ is greater than the initial residue, but it will increase only in proportion to the amount by which the reduction in production of $B$ exceeds the residue.

(c) If equilibrium is initially established in correspondence with a quantity of $B$ that is smaller than, and a quantity of $A$ that is exactly equal to, the respective quantities of maximum partial return, the introduction of the tax will reduce sales of *A* but leave global production unchanged. Hence there will be a residue of *A* and the price of *A* will rise while that of *B* will remain unchanged.

(d) If equilibrium is initially established in correspondence with a quantity of *B* lower than that of the maximum return on this good but with a residue of *A*, the introduction of the tax will as in the previous case reduce sales of *A* without however reducing global production. Hence the residue of *A* will increase and the price of *A* will rise but that of *B* will remain unchanged.

*III. A tax proportional to the quantity of B sold.* A tax of this kind produces effects similar to those indicated above in relation to a tax proportional to the quantity of *A*. The reader will therefore easily be able to work out the individual results relative to the various subcases if he again follows the approach used above in respect of the tax on *A*, applying to good *A* what is there applied to good *B* and vice versa.

*IV. A tax proportional to the quantities of A and B.* A tax applied simultaneously to good *A* and good *B* reduces for both products the quantity corresponding to the quantity of maximum respective return, i.e. it reduces the steepness in the rising phase of the two curves of partial returns. As a general rule therefore we shall see a reduction in global production and an increase in the prices of the two goods. But in the various cases the absolute and comparative rises in the two prices will be different depending on the influence of the other conditions. First of all it is obvious that the amount of the rise in *A* and *B* is influenced by the unitary amount of the tax and in particular by the fact that this is higher for good *A* than for good *B* or vice versa. In the second place the elasticity of the respective demands has an influence. And the way in which equilibrium is ultimately reestablished is influenced by the nature of the initial equilibrium position and by the shift that occurs for both goods in the respective point of maximum return, a shift that makes the ratio *h* higher than, lower than, or equal to the initial ratio after the double tax, depending on whether the shift itself is proportionally greater for *A* than for *B* or vice versa, or on whether it is exactly proportional for both products. Each of these factors acts both on its own account and at the same time together with the others, so the ways in which each factor can combine with the others are different. The definitive result – i.e. the rise in the price of *A* as compared with that of *B* – differs depending on how one factor combines with the others. For one sort of combination makes the rise in the price of *A* greater than that in the price of *B* while another sort makes it smaller. So in addition to the possible combinations that are presented by the solution to the problem and that depend on the initial equilibrium position, there are also combinations deriving from the way in which in each individual case each one of the factors in question coordinates and associates with the others. The problem therefore presents as many solutions as there are possible combinations, but for brevity's sake I shall omit this new series of combinations, which the reader is of course at liberty to study at his own leisure. We need only follow the pattern provided in our previous discussion, taking account of the way in which individual factors combine together case by case.

From this discussion we see that the transposition of a tax, a very straightforward and practically uniform matter in the various cases in a free market, is much more complex under a monopoly regime with substantial differences in its consequences from case to case. The weak and variable correlation between prices of different goods that is a feature of the monopoly regime has its due effect on the transposition of taxes.

# 3 The Theory of Supply at Joint Cost: The Case of Linked Markets

## I A THEORETICAL DISCUSSION OF THE MARGINAL UTILITY OF MONEY

**1.** Up to now I have been studying the laws of joint supply in relation to a closed market and I must now consider these laws in relation to linked markets. To this end I shall follow the pattern used in the first approximation and divide my study into two parts: a first part dealing with the case of free competition and a second dealing with the case of a monopoly.

But before we put the problem on this new basis, a difficulty looms that requires a speedy resolution. In studying the hypothetical case of a perfectly closed market I have been operating with curves of demand and supply which I have taken to be fundamental data of the problem, treating them as basic uncomplicated curves. But this is not in fact the case. As we know, demand curves derive from the combination and fusion of the utility curves of each good and from the utility curve of money. Hence it is legitimate to consider demand curves as fundamental data of the problem and therefore as not being subject to changes in the transition of the market from one equilibrium position to another, but only insofar as it is legitimate to consider the utility curves of the various goods and the marginal utility of money as fixed and constant. Now it is precisely here that doubts arise. Leaving aside for the moment any consideration of the marginal utility of each good, which I shall continue to treat as independent of that of other goods until I am able to deal with it more fully, the hypothesis that the marginal utility of money remains constant – acceptable in the context of a broad approximation in the case of a closed market – seems *a priori* at least to be arbitrary and unacceptable in the case of linked markets. Before proceeding any further it is therefore necessary to clear up this point, so we need first of all to discuss what exactly the marginal utility of money amounts to, for up to now it has been repeatedly referred to without any explanation. We need to clarify how it is computed and the elements of which it is constituted, and see whether on the basis of these constitutive elements it can or cannot be considered constant *grosso modo* in the case of linked markets too.

**2.** As long as we limit ourselves to single individual economies there is no difficulty about how to determine the marginal utility of money. For each individual attributes to each monetary unit a level of utility corresponding to the marginal utility of the goods that may be acquired by it – hence the marginal utility of money is comparable to the marginal utility of the goods that money serves to procure. But the difficulty seems to arise when we move from single individual economies to the economy of an entire market. In the context of an entire market the monetary mass takes on the appearance of a *sui generis* good that will exercise the function of general intermediary in exchange, rather than that of an indirect means for the satisfaction of needs, and this function seems to shift the problem away from its initial basis and make its solution impossible on that basis. But in actual fact this is not the case, and to prove it I need only reduce the problem to the simplest conditions. Let us therefore suppose:

(1) that the good for monetary use does not have any direct utility;
(2) that the quantity of this good may not be increased;
(3) that the development of credit is nil;
(4) that in the unit of time in which exchange takes place the velocity of circulation of money is equal to one.

Given these conditions all difficulties are eliminated. In any market each good has a determinate utility curve arising from the combination and fusion of the various curves of individual utility. Therefore at any moment each good possesses a given marginal utility depending on the nature of its own curve and on the quantity of the good available on the market. Over against this mass of goods there is a determinate mass of money to facilitate their exchange. How is its marginal utility determined?

The function of intermediary in exchange exercised by money in relation to the complex activity of a market does not prejudice its nature as an instrumental good. If in fact it is true that instrumental goods are by nature goods that have no direct utility but that are capable of procuring immediate or complementary goods,[1] no good possesses the characteristics of instrumental goods more than money, at least under the hypothetical conditions we have supposed. It serves to obtain all sorts of goods precisely because it is a general intermediary in exchange, and its instrumentality is to be understood in this sense. Now instrumental goods, being without direct utility, have a mere reflected utility deriving from the utility of the direct goods that can be obtained by means of them.[2] Hence money, as an instrumental good *par excellence*, acquires its own utility from the utility of all the goods it serves to acquire.[3] And each monetary unit therefore has a utility corresponding to the marginal utility of the goods or quota part of goods for which it is exchanged. But each monetary unit may be used equally for the acquisition of any of the various units composing the mass of a given good. These various units, being equivalents, all have the same level of utility, and the utility of each is equal to the utility of the one that functions as a marginal unit.[4] Thus each monetary unit employed for the acquisition of a given good, whatever the unit or part of a unit of this good that is received in exchange, procures an identical sum of utility. But there is more. Since under equilibrium conditions goods are exchanged in quantities inversely proportional to their marginal utility,[5] the marginal quantities of those goods that are exchanged for each other or that are exchanged for a third good, money, are all endowed with the same sum of utility. Thus whatever the category of goods and whatever the unit or part of a unit of each good for which each monetary unit is exchanged, it procures the same utility. This is precisely the marginal utility of money I have been wanting to define. But how to calculate it? If we go on with our hypothesis of a market in which the development of credit is nil, in which all money is actually in circulation, and in which the velocity of circulation of this money is equal to one, we see that in the unit of time under consideration the entire monetary mass is exchanged for the totality of circulating goods. Now if it is true that the utility of money is equal to the utility of the goods that it procures; if it is true that each monetary unit may be applied equally to any of so many units of the various goods, and that each unit can function as a marginal unit and therefore has a utility corresponding to the utility of the latter; finally if it is true that in the unit of time the totality of monetary units is exchanged for the totality of individual units of circulating goods, it follows that the overall utility of the monetary mass[6] is equal to the sum of the utilities of each unit of these goods, their utilities being equal to the utilities of the respective marginal units;[7] in other words it is equal to the sum of the various products obtained by multiplying the marginal level of utility of each good by the number of respective units.[8] Having discovered the overall utility of the monetary mass, it is easy to ascertain the unitary utility from it, i.e. the marginal utility we are looking for: all we need to do is divide the overall utility by the number of monetary units.

With the concept of the marginal utility of money fixed in this way the factors that together determine it are clarified at the same time. It is in fact a function of three groups of variables: (a) the available quantities of each good; (b) the form of the function of the utility of each good; (c) the quantity of monetary units. Let us use $f_0(x), \phi_0(y)$ for the marginal utilities of quantities $x, y$ of goods $A, B \ldots$ and $M$ for the number of monetary units, and we can then represent the marginal utility of money in the following formula:

$$\frac{xf_0(x) + y\phi_0(y) + \ldots}{M}$$

From this it is clear that if all other conditions remain the same the marginal utility of money alters in inverse proportion to its quantity. Since the value of each good corresponds to its marginal utility, if all other conditions remain the same the value of money varies in inverse proportion to its quantity. Hence my conclusions coincide with the conclusions of the quantity theory, a fact which is a testimony to the correctness of these conclusions.[9]

**3.** These conclusions, of capital importance for the theory of the value of money, are of equally great importance for the problem that concerns us here. As we have seen there are three groups of factors that determine the marginal utility of money, but the force with which each one of these factors impacts on the latter is different.

With regard to the first group it should be noted first of all that if for example quantity $x$ of good A increases, the value of $f_0(x)$ or the marginal utility of $A$ decreases. Hence the new value of $xf_0(x)$ will be greater or less than the previous value depending on whether the elasticity of the utility curve is less or greater than one. If the value of $xf_0(x)$ changes then the marginal utility of money necessarily changes too. But first of all for the reasons here set out the change in $xf_0(x)$ – in whatever direction – is always less than proportional to the change in $x$. In the second place the value of $xf_0(x)$ is only one of the elements in the numerator of the expression under discussion. Hence the change in the value of the numerator and therefore of the entire expression is less than proportional to the change in $xf_0(x)$, and since the value of $xf_0(x)$ varies independently less than proportionally to $x$, the change in the available quantity of a circulating good has a fairly minor effect on the marginal utility of money.

As for changes in the form of the utility functions of goods, they have little impact. If, for example, the marginal utility of $A$ increases, the value of $f_0(x)$ and therefore of $xf_0$ increases. But since the numerator of the expression is constituted by other factors which are the more numerous as the circulating goods are more numerous, in this case too the relative change in the value of the overall expression is proportionately less than the change in the factor that has determined it. So changes in the form of the utility function do not bring about any significant changes in the level of the marginal utility of money.

The case is very much otherwise with the quantity of the circulating medium, for we have already seen that changes here bring about inverse changes that are exactly proportional to the marginal utility of money. So of the three possible categories of disturbance this is the only one that has a forceful impact on the marginal utility of money.

From all this it follows that though in considering the marginal utility of money it is certainly legitimate as a broad approximation to ignore disturbances of the first two factors and treat them as hypothetically constant, this is certainly neither legitimate nor possible in respect of the third factor. The impact of disturbances here is so formidable that it cannot be ignored under any circumstances. The question could only be ignored and the marginal utility of money considered as constant if the quantity of the circulating medium remained constant, but while this is hypothetically acceptable in the case of a closed market in which the money stock is not vulnerable to external influence, it is certainly not acceptable in the case of linked markets between which money transfers take place at times of disturbance for reasons of equilibrium. Thus if in relation to the conditions of a closed market it is legitimate (though not strictly rigorous) to consider the marginal utility of money as constant and to continue operating directly with curves of demand, this is not at all legitimate in the case of linked markets even in the context of a broad approximation. So in proceeding to study this case we shall no longer be able to operate with straightforward demand curves but instead in order to represent the individual laws of demand we shall have to take account at the same time both of the marginal utility of each good and of the marginal utility of money. However on the basis of what has been said so far this does not present any obvious difficulty.

Once the marginal utility of money has been discovered by the process indicated above it is easy to make it yield the law of demand for each particular good. All that is needed is to divide the marginal utility of each good corresponding to the quantity of that good available by the marginal utility of money, or multiply it by the inverse of the said utility. If we apply these general norms to individual concrete cases the law of demand for good $A$ is expressed by the following formula:

$$\frac{f_0(x)M}{xf_0(x)+y\phi_0(y)+\ldots}$$

Formulae like these are valid for all other goods, and from now on I shall employ these complex formulae to represent the law of demand for each good in place of the simple functions used hitherto. So we can now proceed directly to the study of our problem and tackle the question of supply at joint cost in the case of linked markets.

## II SUPPLY AT JOINT COST IN THE CASE OF LINKED MARKETS UNDER A REGIME OF FREE COMPETITION

**4.** On the hypothesis of linked markets supply at joint cost may take place both under a regime of free competition and under a monopoly regime, and I shall study the two cases separately with the former as the subject of the present paragraph and the latter of the next. In both cases two problems confront the student. The first problem concerns the way in which where there is supply at joint cost equilibrium initially adjusts between markets that become linked. The second concerns the way in which international equilibrium is temporarily reestablished after any kind of disturbance. This second problem will lead to a study of disturbances of production and the commercial and monetary disturbances that arise from them, and the discussions it will involve will help to clarify this fundamental point: whether and within what limits the presence of supply at joint cost modifies the laws of international monetary circulation, which classical economists have expounded on the basis of the hypothesis of independent costs. All these problems will be discussed and resolved in the pages that follow.

**5.** To this end I shall begin with the straightforward case of just two countries that produce just two goods, goods that are produced jointly. Let $S$ and $T$ be the two countries in question and $A$ and $B$ the two goods produced jointly in $S$ and $T$. Let us suppose that to begin with the two countries constitute two closed markets, and that their respective circulating medium, in each country having a velocity of circulation equal to one, is hypothetically constituted by a good lacking in any utility of its own.

| | |
|---|---|
| With respect to country $S$, let | With respect to country $T$, let |
| – $x_s$ be the quantity of good A produced; | – $x_t$ be the quantity of good $A$ produced; |
| – $f_s(x_s)$ the respective utility curve; | – $f_t(x_t)$ the respective utility curve; |
| – $y_s$ the quantity of good $B$ produced; | – $y_t$ the quantity of good $B$ produced; |
| – $\phi_s(y_s)$ the respective utility curve; | – $\phi_t(y_t)$ the respective utility curve; |
| – $F_s(x_s + y_s)$ the curve of production costs of the two joint goods;[10] | – $F_t(x_t + y_t)$ the curve of production costs of the two joint goods;[10] |
| – $M_s$ the quantity of money in circulation; | – $M_t$ the quantity of money in circulation; |
| – $k_s$ the ratio in which the two goods are produced | – $k_t$ the ratio in which the two goods are produced |

Since the two countries are initially isolated from each other, equilibrium will be established in each of them independently, according to the laws expounded and studied in Chapter 2, part I. These laws, in which the simple demand functions employed up to now are replaced by the complex demand functions designated above, are expressed by the two following systems of equations.

**Equation for Country $S$[11]**

$$\frac{x_{s0}}{x_{s0}+y_{s0}} \cdot \frac{f_s(x_{s0})M_s}{x_{s0}f_s(x_{s0})+y_{s0}\phi_s(y_{s0})} + \frac{y_{s0}}{x_{s0}+y_{s0}} \times \times \frac{\phi_s(y_{s0})M_s}{x_{s0}f_s(x_{s0})+y_{s0}\phi_s(y_{s0})} = \frac{F_s(x_{s0}+y_{s0})M_s}{x_{s0}f_s(x_{s0})+y_{s0}\phi_s(y_{s0})} \tag{1}$$

$$\frac{x_{s0}}{y_{s0}} = k_s \tag{2}$$

**Equation for Country $T$**

$$\frac{x_{t0}}{x_{t0}+y_{t0}} \cdot \frac{f_t(x_{t0})M_t}{x_{t0}f_t(x_{t0})+y_{t0}\phi_t(y_{t0})} + \frac{y_{t0}}{x_{t0}+y_{t0}} \times \times \frac{\phi_t(y_{t0})M_t}{x_{t0}f_t(x_{t0})+y_{t0}\phi_t(y_{t0})} = \frac{F_t(x_{t0}+y_{t0})M_t}{x_{t0}f_t(x_{t0})+y_{t0}\phi_t(y_{t0})} \tag{3}$$

$$\frac{x_{t0}}{y_{t0}} = k_t \tag{4}$$

Given these two systems of equations, let us now suppose that the two countries enter unexpectedly into commercial relations, and let us suppose moreover that initially the prices of the two goods in the two countries are such as to generate imports of good *A* from *T* into *S* and imports of good *B* from *S* into *T*. If, as is likely, the value of the imports of good *A* into *S* is not exactly equal to the value of the exports of good *B* from *S* there will then be a transfer of money from one country to the other. Let us then suppose that the relative value of the imports is such as to require the transfer of a given quantity $M_1$ of currency from country *S* to country *T*.[12]

As soon as the two countries enter into commercial relations a general change will take place in the distribution of the circulating medium, the distribution of the production of the two goods between the two countries, their cost, and their respective prices. International equilibrium will be established as soon as the prices of goods *A* and *B* are on a level in the two countries and as soon as in each of them the global price coincides once again with the respective global cost.[13]

On this basis, and assuming that the two countries *S* and *T* are already in equilibrium, let us see what disturbances derive from a change in the principal equilibrium conditions, and in what way where there is supply at joint cost international economic equilibrium is reestablished if disturbed.

The disturbances to be considered here are as follows:

1. An increase in demand for good *A* in country *S*;
2. A fall in the cost of production in country *S*;
3. The introduction of a duty on imports of *A* into *S*;
4. The introduction of a premium on exports of *A* to *S*.

## 6. Case 1 An Increase in Demand for Good *A* in Country *S*

A disturbance of this kind resolves as we know into an increase in successive demand prices corresponding to the successive quantities demanded. Now if we ignore for the moment possible changes in the marginal value of money, the increase in demand prices can only derive from an increase in the marginal utility of good *A*. Hence, if we are operating with utility curves, as is the case here, the increase in the marginal utility of *A* may be considered as the factor determining the increase in demand. But immediately we see that as the value of $f_s(x_{s_o})$ increases, the value of the expression $x_{s_o} f_s(x_{s_o})$ increases too. This results in an increase in the marginal utility of money and means that the rise in the demand price for *A* is less than proportional to the increase in the marginal utility of *A*. And whatever the actual extent of the disturbance it is certain that an increase in the marginal utility under discussion causes a rise in demand for *A* which raises the price of *A* and results in an expansion in global production in *S* and consequently a fall in the price of good *B*.[14]

This swift and automatic fall in *B* introduces a strange new factor into the process of the recovery of equilibrium, notably affecting two crucial areas – the extent of the overall disturbance and international money circulation.

Where the first of these areas is concerned it is evident that once disturbed by a rise in demand for *A* in *S* international economic equilibrium is reestablished as soon as the global price in the two countries again coincides with the respective global cost. Now the fall in the price of *B* that occurs in *S* with the rise in the price of *A* means that if all other conditions remain the same the adjustment of global price to global cost happens in *S* through a rise in global production and therefore in the production of *A* that is less than in the case where the production of *A* is independent, and less in proportion as the fall in the price of *B* is more substantial, i.e. in proportion as the elasticity of demand for this good is smaller. But the rise in the price of *A* in *S* brings about its import from *T*. And in *T* the export of *A* raises its price, encourages global production and causes *B* to go up. So in country *T* too the adjustment of global price to cost occurs through an increase in production and therefore in exports of *A* that is less than in the case of independent costs, and less to the extent that the fall in the price of *B* is more substantial. Hence overall the disturbance in global production and in the export of *A* deriving from the increase in demand for this good (assuming parity in that increase) is smaller in the case of joint costs than in the case of independent costs.

Where the impact on international money circulation is concerned, there are certain interesting and strange aspects that need to be discussed. The automatic fall in the price of *B* – the consistent basis of the whole process we are studying – opens up the possibility that country *S* may export the said good immediately. This in turn seems to offer a way for country *S* at least partially to finance imports of *A* with an obvious saving in the transmission of money. In this second area then our task is to consider in what cases the export of *B* occurs and to establish whether as this export takes place it does or does not really modify the laws of international money circulation.

In country *S* the rise in demand for *A* causes a rise in the price of *A* and thus as we have seen brings about an increase in global production and a fall in the price of *B*. But since the rise in the price of *A* in *S* brings about the import of *A* from *T* and a rise in its price in *T*, the same things will happen in *T* as in *S*. Ultimately therefore the new equilibrium will be established through an increase in global production in *S* and in *T*, a rise in the prices of *A* and a fall in the prices of *B* in both countries. The proportion in which each country shares in the overall rise in global production will depend on the curves of costs and on the elasticity of demand for the two goods in the two countries. And whether or not the immediate export of good *B* from country *S* can or cannot take place will depend on the relative elasticity of the curves of demand for it in the two countries.[15] We shall study the main cases that can occur. To make our task easier let us suppose that in the two countries the global cost curves and demand curves for *A* are equal in all cases, and let us also suppose that the demand curve for good B varies from case to case in *S*. There are then three possible combinations of these conditions; (a) the elasticity of demand for *B* in *S* is equal to that in country *T*; (b) it is smaller; (c) it is greater. Let us look at each of these three cases separately.

(a) In this first case the rise in global production resulting from the growth in demand for *A* in *S* takes place in the same proportions in the two countries, which means that for each successive hike of production in *S* there is a simultaneous equal hike in *T*. But in addition – and this is a point of capital importance – as global production rises in the two countries the fall in the price of *B* in *S* is matched by an equal fall of *B* in *T*. This initially makes the export of *B* from *S* to *T* impossible. But as *S* imports good *A* from *T*, it has to export money in exchange, and after this export has led to a fall in all prices in *S* good *B* – now priced lower in *S* than in *T* – then becomes exportable and is in fact exported. Hence in this case equilibrium is recovered through the import of *A* into *S*, the export of money, and the subsequent export of *B* from *S*; i.e. it is recovered through the very same process that operates where there are independent costs. In the case under consideration therefore the existence of joint costs does not in any way modify the established laws of international monetary equilibrium.

(b) In order to examine the second case let us now suppose that with all other conditions remaining unchanged demand for *B* is less elastic in *S* than in *T*. Given these conditions each successive rise in global production in the two countries leads to a fall in the price of *B* that is of greater proportions in *S* than in *T*, so it becomes advantageous for *S* to export good *B*. Exports of *B* take place immediately and this leads to some rather interesting consequences. For while in the preceding case and in all cases of production at independent cost, imports of *A* are matched by exports of *B* only after a transfer of money has taken place from *S* to *T*, in the present case exports of *B* take place at least in part even before the outflow of money occurs. So here production at joint cost genuinely modifies in certain respects the laws of international money circulation as we have so far understood them. However these are modifications that alter certain external aspects of the fundamental process without affecting its innermost nature. In this case, unlike in the other cases, imports of *A* are accompanied by immediate exports of *B* which to some extent serve as payment for them. But these exports do not therefore reduce the extent of the prevailing disequilibrium of the money balance (as might have been anticipated), still less do they avoid direct international money transfers, generally necessary to settle that balance definitively. To satisfy ourselves of this it is sufficient to compare the specific conditions of this case with those of case (a).

As we know, equilibrium is established in all cases as soon as individual prices in the two countries are on a level and respective global prices coincide with respective global costs. Now in the present case, precisely because the fall in the price of *B* is greater in *S* than in *T*, for each successive increase in global production *B* goes down more in *S* than in *T*, which means that in spite of the fact that the respective curves of costs and the other conditions in the two countries are identical the coincidence of price with global cost is reached in *S* with a smaller increase in global production than in *T*. If the increase in production cannot as in the previous case be equal in the two countries, but must be smaller in *S* than in *T*, it will necessarily be smaller in *S* and greater in *T* than it was in the preceding case.[16]

Now if the increase in global production is smaller in *S*, the increase in production of *A* is also necessarily smaller. And if domestic production of *A* is smaller than in the preceding case imports of it must be more extensive if equilibrium is to be recovered. Hence the rapid fall in the price of *B* first of all leads to more substantial imports of *A*. The automatic export of *B* takes place as a consequence of the rapid fall in the price of *B*, and the more rapid this fall the more the good becomes exportable. But on the other hand since a greater fall in the price of *B* results in a smaller rise in production it also results in less extensive potential exports. Therefore while the rapid fall in the price of *B* automatically triggers exports of *B*, it also limits the extent of such exports; in other words it ensures that automatic exports of *B* cannot overall be very substantial under any circumstances.

Therefore while the rapid fall in the price of *B* in *S* on the one hand encourages the automatic but limited export of *B*, on the other hand that fall gives rise to more extensive imports of *A* by comparison with the preceding case. At the same time that it provides for the settlement of a part of the imports it also tends to increase them. Its effects on the money balance cancel each other out because they are contradictory. Hence after the imports of *A* and the automatic exports of *B* a deficit in the money balance still remains to be settled, and this deficit, similar in amount to the one in the preceding case, obviously has to be settled in the same way. Thus in this case the process of recovery of equilibrium considered step by step in its successive phases operates as follows. Imports of *A* more substantial than in the preceding case are immediately followed by automatic exports of *B*, but as this happens only in proportion to the excess of imports of *A*, it leaves a part of the money balance still to be settled.

As in the preceding case this settlement takes place through the export of money from *S* and the subsequent export of *B*. So while the automatic export of *B* introduces a new element of international settlement, it does not in fact have the power to limit or restrict the subsequent transfer of money, for the balancing action that it seems to have is neutralised by an opposite effect and loses any real consistency. All things considered the only difference in the final result is that in this case equilibrium is recovered by imports of *A* and exports of *B* which are both greater overall than in the preceding case.

(c) Let us now move on to the final case and suppose that with all other conditions remaining the same, demand for *B* is more elastic in *S* than in *T*. Under such conditions with each successive increase in global production in the two countries brought about by the increase in demand for *A* in *S*, the price of *B* falls more rapidly in *T* than in *S*; i.e. by making it advantageous for *T* to export good *B* it brings about the immediate export of that good. In this case too therefore there is an automatic transfer of good *B*, but instead of taking the opposite direction to the transfer of *A* it takes the same direction, thus accentuating rather than countering the disturbing effects. So the imports of *A* and *B* together help to create the overall disequilibrium of the money balance, and in this case we have the singular anomaly that the increase in demand for a given good in a given country gives rise not just to imports of this good but also to imports of the other joint good or goods.

But though the automatic imports of *B* work cumulatively together in this case with those of *A* to create the overall initial disequilibrium of the money balance, the latter is not greater (assuming parity of other conditions) than in cases (a) and (b). Here again we are dealing with one of those disturbances that set in motion compensatory forces that will eliminate their effects. The rapid fall in the price of good *B* in *T* triggers exports of *B*; but as the fall in the global price speeds up it leads to a coincidence of this with the global cost – i.e. definitive equilibrium – being reached with a smaller increase in production in *T* than in *S* and consequently a smaller increase than in the two preceding cases. In this case the increase in global production in *T* is smaller and therefore of course the possible export of *A* from *T* is smaller too so while country *T* automatically exports good *B* its exports of *A* are less substantial. The automatic export of the former good takes place at the expense of the export of the latter, so all in all the actual disequilibrium of the money balance is more or less the same overall as it was in the preceding cases. In this case too then while supply at joint cost brings about the automatic export of good *B* by *T* it does also in certain respects modify the process through which equilibrium is recovered, without extending the overall disturbance and without doing anything (as one might have expected) to make the money transfer necessary for the recovery of equilibrium more substantial than in the preceding cases.

To summarise then, there are two kinds of influence exerted by supply at joint cost on international economic equilibrium when there is an increase in demand for a joint good in a country. First of all it reduces the overall extent of commercial and monetary disequilibrium by comparison with what occurs (assuming parity in the increase of demand and all other conditions) in the case of independent costs. This effect it has in all cases without distinction, but at the same time it exerts another sort of influence, which relates exclusively to the process by which equilibrium is recovered. This second sort of influence it exerts first of all only in given cases and conditions and it does so moreover in such a way that it counterbalances the various disturbing elements that it produces in the different cases with equivalent compensatory elements. So in spite of the different chain of actions and reactions that result in the different cases, the overall amount of the deficit in the money balance is virtually the same in all cases. Hence this second sort of influence, unlike the first, has not the least effect on the intensity of the overall disequilibrium.

### Case 2 A Fall in the Global Cost of Production in Country *S*

As a result of technical progress the global cost of production falls in *S* while all other conditions remain the same. Production of *A* and *B* in *S* will grow and the price of both products will fall, making it profitable to export them.

Imports of these products into *T* will drive their price down and reduce production, leading to the export of money from *T* at the same time. In the final analysis the fall in the cost will have the effect of increasing the money stock of the country in which it has occurred. In this respect the presence of supply at joint cost in no way modifies the well known effects of technical progress in one branch of production where there are independent costs.[17]

### Case 3 The Introduction of a Duty on Imports of *A* into *S*

Referring back to our initial example, let us suppose as we did then[18] that as long as they are isolated from each other conditions of domestic demand and of costs in the two countries *S* and *T* are such that when the countries enter into commercial relations country *S* imports good *A* and exports good *B*. Let us now suppose that country *S* introduces a duty on imports of *A*. Since in equilibrium conditions and in the absence of duties prices in the two countries tend to be at the same level; since when this happens imports and exports reduce to zero; and since finally where commercial movement is nil the mere introduction of a duty cannot of itself disturb an established equilibrium, we cannot as we have done up to now study the effects of such a duty by starting from an equilibrium position already established and positing the onset of a disturbance; instead we must posit an existing initial disequilibrium and see what is the difference between the two equilibrium positions to which the market moves with and without the duty in question.

When put in this way the problem is easy to solve. To begin with one consequence of the duty will clearly be that the levelling of the price of *A* between the two countries will be only partial. After the two countries have developed links the price of *A* in *S* will fall less than it would have done in the absence of the duty, imports of *A* into *S* will be less substantial, but domestic production in *S* will be more substantial. But in producing a greater quantity of *A* country *S* will also produce a greater quantity of *B*, so the price of *B* will fall in *S* and exports of this good will be more substantial than what they would have been in the absence of the duty. In country *T* exports of good *B* will be smaller and imports of good *B* greater than what they would otherwise have been. Equilibrium will be established as soon as in the two countries prices of *B* are exactly the same, prices of *A* differ by the exact amount of the duty, and global prices are exactly equal to respective global costs.

Much more interesting is the way the duty affects the monetary balance and the orientation of international trade. By making imports of *A* into *S* smaller and exports of *B* from *S* more extensive than they would otherwise have been, the duty increases imports of money into *S* or reduces any exports of it that would have taken place in the absence of the duty. Thus the duty has the effect of increasing the monetary endowment of the country that establishes it, in other words its impact in the case of joint cost is identical with what we have seen in the case of independent costs.[19]

There is however a difference between the two cases with respect to the way in which the duty operates. While in the case of independent costs it merely reduces imports of the protected good, in the case of joint costs it not only limits imports of this good but encourages the export of the other. Here then we have a second point of capital importance. While on the one hand the duty gives effective protection to the good in favour of which it has been imposed, on the other hand it acts as a premium on the export of the joint good, and unlike in the case of independent costs, its disturbing effect thus operates in two directions.

But though there are two directions there is not a double force, for this would require that the duty first of all have as powerful a protective effect in favour of *A* as in the case of independent costs, and then that it encourage the export of *B* as well, but this does not happen. Both in the case of independent costs and in the case of joint costs the duty tends to encourage the production of the protected good. Where the duty is the same and all other conditions unchanged its protective action is the more powerful as the increase in production it triggers is greater. Now the duty achieves this protective effect through a rise in the price of the good it is intended to protect, while the rise in price is also what enables it to boost production. Production is increased as soon as the price rises above the cost as a result of the duty and it goes on increasing until through a series of convergent movements price and cost eventually come to coincide once more. The actual increase in production necessary for the adjustment of price to cost depends therefore on the curve of costs and the curve of demand for the good under consideration, so if all other conditions are the same the protective power of the duty depends on these two curves.

To simplify the discussion we shall suppose that the curve of costs is identical in all cases. On this hypothesis the increase in production remains entirely subordinate to the curve of demand, i.e. to the curve of prices. Now it is precisely in relation to this that arise the differences between the case of independent costs and the case of joint costs. The more the price rises initially as a consequence of the duty, the greater the increase in production necessary (with all other conditions remaining the same) for price and cost to coincide once again, but while in the case of independent costs it is the rise in the price of *A* that has a decisive impact, in the case of joint costs it is the rise in the global price. Now let us suppose that in the two cases there is an identical duty on imports of *A*. In the case of independent costs the price of *A* initially rises by the entire unitary amount of the duty. The same thing applies in the case of joint costs. But since the global price derives from the combination of a fraction of the price of *A* and a fraction of the price of *B*, the rise in the global price will be less than the rise in the price of *A*, i.e. it will ultimately be less than the amount of the duty. Hence if the duty is the same, the price that influences the subsequent increase in production rises in greater proportions in the case of independent costs than in the case of joint costs. If the rise in the said price is smaller, the increase in production that derives from it is consequently smaller and hence the protective action of the duty is weaker too, if all other conditions remain the same. But the duty also encourages export of good *B* and it achieves these two results simultaneously. Thus the function of premium on export exercised by the protective duty in the case of joint costs is fulfilled at the expense of and to the detriment of its protective function.

This conclusion is reinforced by other important considerations. In the case of independent costs the duty only encourages the production of *A* and the speed with which the price tends to fall with each successive increase in production depends purely on the elasticity of the curve of *A*. But in the case of joint costs the duty encourages the production of *B* as well as that of *A* and the speed with which the global price falls depends on the elasticity both of the curve of demand for *A* and of the curve of demand for *B*. If the elasticity of the curve of the two goods is such that their prices fall in the same proportions, the reduction in the global price will proceed in parallel with the fall in the price of *A*, and the fall in the price of *B* will not have any later repercussions. But as the rise in the global price resulting from the duty is smaller than the rise in the price of *A*, the export of *B* will take place at the expense of the protection of *A*. If the elasticity of the curve of the two goods is such that the price of *B* falls more quickly than that of *A*, the rapid fall in *B* will facilitate its export, so accentuating the effect of the duty as a premium on export. But at the same time this rapid fall in *B* speeds up the fall in the global price and also speeds up the adequation of the latter to cost, i.e. causes this adequation to come about through an increase in production that is smaller than on the previous hypothesis. If the elasticity of the curve of the two goods is such that the price of *B* falls less rapidly than that of *A* the slow fall in *B*, hampering its export, will weaken the effect of the duty as an export premium. But at the same time the slow fall in *B*, putting a brake on the fall in the global price, will slow down the adequation of the latter to the cost, i.e. will bring about this adequation through an increase in production that will be greater than on the first hypothesis and consequently greater than on the second. Hence the two functions of the duty are always at odds with each other. So in the different cases the protective action is intensified by as much as the exporting action is attenuated and vice versa. The duty cannot increase the force of one of these two actions without reducing that of the other; it cannot exercise its secondary function without damaging the effectiveness of its main function. Hence while this again confirms the principle that the exporting function of the duty is accomplished at the expense of its protective function, it also demonstrates that where there are joint costs the twofold function exercised by the protective duty does modify the way in which that duty disturbs the monetary balance of the country that imposes it but this function does nothing to increase the force of the monetary disturbance that results. Therefore this disturbance may be considered overall as almost analogous to or at least not greater than the one that occurs in the case of independent costs.

These conclusions suggest a final consideration. The fact that the protective role of the duty is smaller in the case of joint costs than in the other case shows that the presence of joint costs has a damaging effect on it, and the research we have done on the reason for this throws fresh light on this important problem. Up to now we have been considering the case of just two goods produced jointly, but there are cases in which there are more than two joint goods. A study of the composition of the global price in these cases reveals that the more numerous the joint goods the smaller the fraction of the price of each good that participates in the formation of the global price, and therefore the smaller the rise in the global price under the impact of a given increase in the price of one of such goods. If therefore at a given moment a protective duty is imposed on one of the joint goods, the price of it will initially rise by the entire amount of the duty, but the global price will rise in smaller proportions to the extent that the number of joint goods is greater. Thus the greater this number the smaller the impulsion to increase production and the smaller the actual increase in production of the protected good. Hence we conclude that the greater the number of goods produced jointly the weaker the protective effect of a duty established in favour of one of them.

### Case 4 The Introduction of an Export Premium on *A* into *S*

Let us suppose that between countries *S* and *T* there is a perfect economic equilibrium, i.e. there are no transfers of goods or money, and let us suppose that all at once country *S* puts a premium on the export of *A*. What will be the result?

Good *A* will immediately be exported by *S* to *T* so that its price will rise in *S* and its production will increase, but at the same time production of *B* will also increase. The price of *B* will fall in *S* and this will encourage its export. *A* and *B* will both be imported into *T* at the same time, the price of both goods will fall and their production will be reduced. Overall production will increase in *S* and decrease in *T* but in *S* production will push up towards a higher level of costs than that from which it is falling in *T*, so the shift of equilibrium will be accompanied by a destruction of wealth. The price of *A* in *S* will be higher than before while prices of *B* in *S* and of *A* and *B* in *T* will be lower, and equilibrium will be established as soon as prices of *B* have levelled out in the two countries, prices of *A* have moved away from the amount of the premium, and global prices in the two countries are exactly equal to respective global costs.

From the fact that the premium on the export of *A* encourages the export of both *A* and *B* from country *S* we draw the general principle that a premium on the export of a good produced jointly with others resolves into a premium on the export of all goods produced jointly with the one carrying the premium.

But here too as in the case of the duty, what the action of the premium gains in extent it loses in force. The effectiveness of a premium on export is measured by the amount of the exports of the product carrying the premium that it generates. This amount is formed by a combination of the quantity by which production rises through the effect of the premium and the quantity by which domestic consumption falls as a consequence of the rise in the price triggered by that premium. Now to be able to establish in which of the two cases overall export is the greater we need only ascertain whether and in what direction the two flows of the protected good that feed its export differ quantitatively in the two cases, and to do this we need only make a separate study of the effects of the premium in the two successive phases when its action is operative.

In fact the premium in question triggers exports of a part of the product already in existence immediately, but only after a certain period of time does it actually boost production. Thus its complete action develops in two distinct phases which to some extent correspond to the two successive phases in which the two flows in question form one after the other.

In the first phase the premium triggers only the first flow – i.e. it generates exports exclusively at the expense of the mass of the product carrying a premium that is already present in the country where the premium is applied. The international market is disturbed by this initial effect of the premium but immediately adjusts to a temporary equilibrium position until growth in production comes on the scene as a new disturbing element. This equilibrium is reached as soon as – on account of the rise in price in the exporting nation and its fall in the importing nation – consumption of the product carrying a premium is reduced in the first country by the precise amount by which it is increased in the other country, and as soon as the prices of the said product are separated in the two countries by the the precise unitary amount of the premium. Now since this adjustment of equilibrium takes place without any increase in production the actual characteristics and the trajectory of production obviously have no influence on it. Whatever be the law to which production conforms or the ratio in which the various goods are produced jointly with the good carrying the premium, equilibrium is established in the same way in all cases (assuming all other conditions are the same) through the same shift in its constitutive elements. Therefore the overall amount of the exports of the product carrying the premium is at this first moment the same whether costs are independent or whether they are joint.

Once this first equilibrium position is reached then the other factor comes into play with a disturbing effect in the second phase. After a certain period of time the rise in price provoked in the first phase by the premium eventually triggers an increase in production. The new mass of products coming on to a market in equilibrium creates a state of dynamic disequilibrium and causes a general readjustment which takes place through further exports of the product carrying the premium. It is at this precise point that the second of the two flows under consideration comes into operation, and it is in relation to this that the differences between the two cases under discussion arise.

In this second phase the export of the product carrying the premium derives from the increase in production and tends to be proportional to the amount of that increase. Now the increase in production depends as we know on the initial rise in the regulatory price, and if all other conditions are the same it is greater to the extent that this rise is the more substantial. The rise in the regulatory price is a response to the rise in the price of *A* triggered by the premium in the first phase, a rise in *A* that we have seen to be identical in the first phase in the two cases of independent costs and joint costs. But while the force of this rise is identical in the two cases the reflexive effect on the regulatory price is different. In the case of independent costs the regulatory price of production is the price of *A* while in the case of joint costs it is the global price. Now in the first case since the price of the product carrying a premium is identical with the regulatory price the rise in the latter corresponds to the rise in *A*. But in the second case for reasons mentioned several times already the rise in the regulatory price – that is in the global price – is smaller than the rise in *A*. If the rise in the regulatory price in the case of independent costs is greater than in the case of joint costs, the increase in production arising from it is greater in the former case. And since the amount exported is proportional to the increase in production, exports of the product carrying a premium are more substantial in this second phase in the case of independent than in the case of joint costs. Now if it is true that in the first phase such exports are identical in the two cases while in the second phase they differ in the direction indicated, it follows that if all other conditions are the same exports of the product carrying a premium are overall greater in the case of independent than in the case of joint costs. This means that the effectiveness of the premium in bringing about the export of the product carrying a premium is greater in the first than in the second case. But as we have seen, in the second case the premium also triggers the export of all goods produced jointly with the good carrying it, so although losing out in terms of force of impact the premium gains in breadth. Therefore the export of the joint goods that the premium under discussion fosters takes place at the expense of and to the detriment of the export of the good carrying a premium. If in the case of joint costs by comparison with the case of independent costs the effect of the premium loses force because it gains in extent, clearly the force of that action is the weaker as its extent is broader. Now the effect of the premium is the greater, the greater the number of goods produced jointly with the good carrying a premium. Thus the greater this number, the less the effectiveness of the premium itself – assuming parity in the premium and in all other conditions. This leaves us with the possibility of an extreme case in which however high it is a premium remains without any notable influence on the export of the good carrying it solely because this good belongs to a very numerous group of joint goods.

## III SUPPLY AT JOINT COST IN THE CASE OF LINKED MARKETS UNDER A MONOPOLY REGIME

**7.** First of all a preliminary observation. A strict and exclusive monopoly in several markets would exclude any possibility of intercommunication between them, so to talk about linked markets under a monopoly regime seems like a contradiction in terms. But there is no contradiction if we simply adopt the hypothesis of a merely partial monopoly, a hypothesis that is all the more acceptable and legitimate where there is supply at joint cost. So in dealing in this chapter with the case of monopoly between linked markets we shall be referring exclusively to the condition of a partial monopoly.

For simplicity's sake I shall continue to suppose that there just two linked countries, *S* and *T*; that there are just two goods produced jointly, goods *A* and *B*; and that these two goods are produced both in *S* and in *T*, and are in fact the only goods that the countries produce and exchange. Given these conditions a partial monopoly may be established in two different ways.

(I) In market *S* there exists a monopoly in the production and sale of *A* and *B* but in market *T* the production and sale of *A* and *B* take place under a regime of free competition both domestically and externally, so that the producer in market *S* can if it suits him send part of his production to *T*. For this to be the case we must assume the existence in market *S* of protective or prohibitive customs tariffs that safeguard the monopoly there for both products.

(II) Each within the limits of its own market, countries *S* and *T* possess a monopoly over production and sale of good *A*, while the two countries are open to each other for the sale of good *B*.

**Case I**

**8.** In this case as we have seen the producer in country *S* has a monopoly of production and sale of the two goods *A* and *B* in the domestic market, and he is also free to export these goods whenever it suits him to the market in country *T*, which according to our hypothesis is open to them. As a monopolist the producer in country *S* will impose monopoly prices on that country's market, but he will not be able to impose such prices on market *T* where free competition prevails and prices have to be adequated to costs so that they are normally lower. If he is to be an exporter whenever conditions are favourable, the producer in country *S* has to charge lower prices abroad than at home,[20] so here exports involve discriminatory prices for the markets of the two countries. The prevalent view is that this discrimination is 'dumping' whenever the producer in country *S* sells in the other country at prices lower than the costs to the producers in that country.[21]

But if this is the necessary condition for the difference in the prices to be regarded as genuine dumping, the possibility of distinguishing ordinary price differentiation from dumping seems to disappear in the case of joint goods, and since it is precisely this case that we have to study here we need to establish immediately whether the difficulty is a real one. In the case of joint costs there is not a specific cost for each of the goods produced jointly and it is this absence that seems to create the difficulty in question. However where the producer in country *S* sells both goods *A* and *B* in *T* at prices lower than the ones initially prevalent there, it is obvious that there is undercutting in *T* and any doubt about the matter is dispelled. But it may happen – and it is in respect of this case that doubts occur – that the producer in country *S* supplies good *A* in *T* at a price lower than the prevalent one there, while deliberately abstaining from supplying good *B* or while supplying it at a price so high as to preclude any likelihood of sales, at least for the moment. In this case, given the impossibility of establishing the specific cost of *A*, the difficulty does seem to be present but in reality it is non-existent here too.

In fact the fundamentally important thing in supply at joint cost is not the individual price of the separate goods but the global price, and in each case it is this that will be taken as the term of comparison *vis-à-vis* the cost. To begin with in country *T* the global price coincides with the marginal global cost, and here there cannot be any question of sale below cost price. But if the producer in country *S* now sells good *A* in country *T* at a price lower than the initial one, this causes both a fall in *A* and a fall in the global price in that country. Since initially the global price coincided with the cost, from this moment it falls below cost. Therefore after the fall in the price of *A* production in country *T* – which at the initial price of *A* was operating at cost – takes place at prices below cost, so that sales of good *A* in country *T* by the country *S* producer at a price lower than the initial one translate for the country *T* producer into sales at a price lower than his own cost.

Therefore once again in the case where of two joint goods only one is sold on a foreign market, it is perfectly possible to determine whether or not it is sold at a price below cost in *T*.[22] Since the prevalence or otherwise of dumping depends on this, international price discrimination pure and simple can always be distinguished from dumping – despite all appearances to the contrary – and even in the case of production at joint cost.

**9.** Export with relative discrimination of prices may have various purposes and operate in different ways depending on those. Here I shall pass over bellicose aims such as the violent conquest of foreign markets or the annihilation of foreign industries and consider only the pacific aim of the immediate increase of the monopolist's profits.[23] In this respect export may take place through an appropriate increase in production or without any such increase, so there are two cases that may present themselves and they need to be studied separately. The second case is paradoxical in appearance and as we shall see peculiar to supply at joint cost, but let us begin with the first.

**10.** With respect to the first case it is obvious that as a monopolist the producer in country *S* will tend to maximise his own net profit and will allow this aim to dictate his line of conduct in regard to the foreign as well as the domestic market. First of all he will strive to maximise his profit from domestic sales, producing the appropriate quantity for this, and he will then proceed to produce for export only to the extent that he manages to get a further increase in profits from it continuing to increase production for export until his net profit reaches the highest level allowed by all the other conditions. Put in this way the problem turns out to be a problem of maximum profit and finding a solution to it involves finding a solution to these two fundamental points, i.e. the discovery of

(a) the conditions requisite for production for export;

(b) the point to which, where the conditions as in (a) are prevalent, the monopolist finds it profitable to extend production for export.

In relation to the first point it is essential to bear constantly in mind that for the producer to produce the supplementary quantity for export he must anticipate an increase in his net profit from it. This as we know results from the difference between the total gross return and the total cost, and while production for export implies an increase in the total return it also involves an increase in total cost. This then will leave a supplementary net profit whenever the increase in the supplementary gross return that it yields is greater than the increase in the total cost involved. But where production for export is concerned we need only suppose that a very tiny quantity of *A* and *B* is produced for this purpose – with even an infinitesimal quantity of *A* and *B* being produced for export the relevant conditions may be considered to apply. The solution to the first point therefore involves ascertaining the conditions in the presence of which the supplementary return deriving from the production of an infinitesimal quantity of *A* and *B* for export is greater than the supplementary cost.

For simplicity's sake we shall suppose that the infinitesimal quantity in question is a unit of global production of *A* and *B*. On this hypothesis the gross supplementary return is represented by the global sale price of the global unit of *A* and *B* in *T*. There will be production for export whenever the said global price is even the smallest amount higher than the supplementary cost it involves. The global sale price comes from the two individual prices of *A* and *B*. Now if we suppose as we have done in other similar cases that the market in *T* is initially a closed market and that it opens to foreign products quite suddenly, we can see that the base prices on which the country *S* producer can count are the prices of *A* and *B* prevailing in *T* before its opening to foreign trade. The prices of these two goods together form the initial global price in *T*. It is therefore enough that the producer in country *S* supply the two goods *A* and *B* in *T* at a global price equal to or, to be more precise, infinitesimally lower than, the initial global price in *T*, for him to succeed in placing the unit of global product in question.[24] Hence the initial global price in *T* may be regarded as and assumed to be the expression and measure of the supplementary return obtainable from production of a unit of the global product for export.

As for the supplementary cost, that depends on the average global unitary cost and on the trajectory of the curve of average global unitary costs in proximity to the amount of global production corresponding to the point of maximum net profit from domestic sales. In proximity to the point indicated the monopolistic company in country *S* can produce at either uniform, rising or falling average global unitary costs depending on its level of development, and on the trajectory of these average costs depends the trajectory of the supplementary cost and consequently in individual cases the solution to the first part of our problem.

If average unitary costs are constant every production increase is accompanied by a proportional rise in total costs, and this means that an increase of one unit of global product involves a supplementary cost exactly equal to the new average global unitary cost. So in this case the supplementary return will be greater than the supplementary cost and there will be production for export whenever the initial global price in country *T* is higher – even if only infinitesimally so – than the average global unitary cost to the country *S* producer corresponding to the extent of production indicated above.

If unitary costs are on the up every production increase is accompanied by a rise in the total cost that is more than proportional to the increase in quantity; this is because in addition to the cost involved in the supplementary quantity there is the increase in the cost of the quantity produced previously. Thus an increase of one unit of global product involves a supplementary cost greater than the new average global unitary cost, and in this case the supplementary return will be greater than the supplementary cost and there will be production for export only to the extent that the initial global price in country *T* is higher than the new average unitary cost to the producer in country *S* – higher than it to a certain minimum degree.[25]

If unitary costs are on their way down any increase in production is accompanied by a rise in the total cost less than proportional to the increase in quantity; this is because the cost involved in the supplementary production is balanced by the fall in the cost of the quantity produced previously. So an increase of one unit of global product involves a supplementary cost lower than the new average global cost and in this case the supplementary return will be greater than the supplementary cost and there will be production for export even if the initial global price in country *T* is lower than the new average global unitary cost to the producer in country *S*, provided it is lower within given limits.

To conclude then, production for export becomes profitable and takes place whenever the initial global price in country *T* is even fractionally higher, or higher beyond given limits, or lower within given limits, than the global unitary cost to the country *S* producer – depending on whether production in *T* takes place at uniform, rising or falling cost.

**11.** Having established the necessary conditions for production for export, we now need to settle the second point, i.e. to establish the extent of such production in the different cases. The country *S* producer can sell different quantities of each of two goods *A* and *B* in country *T*, but gradually increasing quantities are matched by gradually falling sale prices. For each quantity put on sale there is a corresponding determinate price, and the two series of prices at which the successive quantities of the two goods can be sold form the two sale curves for goods *A* and *B* in market *T* for the country *S* producer.

These two curves derive from the combination of the curves of demand and supply in country *T*. The curves of demand establish the total quantities of the two goods respectively demanded at the individual prices. But national producers partially meet the demand that corresponds to each individual price, so the quantity of each good that the country *S* producer can sell in country *T* at each individual price is given by the quantity demanded overall in *T* at that price, *minus* the quantity offered by national producers at that price. So with respect to supply to market *T* we find ourselves in presence of one of those cases that Marshall refers to of rival supplies.[26] But supply by producers from *T*, like the rest of the supply available in country *S*, is a joint supply of the two goods *A* and *B*. Now our earlier research has shown that the quantity of one of the two joint goods that is offered at its various prices depends not only on the curve of global marginal costs but also on the price of the other good. Thus the domestic supply of good *A* at the various prices depends not only on the successive global marginal costs but also on the concomitant prices of *B*, and vice versa the supply of *B* depends on the concomitant prices of *A*. Now since these domestic curves of supply of goods *A* and *B* influence the trajectory of the respective curves of sale of the producer in country *S*, and since each of these curves of supply depends among other things on the concomitant price of the other good, a price which influences the respective curve of supply and therefore the respective curve of sale, these two sale curves must be connected in some way. A change in the shape of one involves a change in the shape of the other. Thus we are faced with a complex case in which two curves of rival supplies combine with two curves of joint supply so that the result depends on the combined action of the one pair with the other.

Now if for the moment we remain with generalities and ignore these complications, it is easy to deal with the overall trajectory of the two sale curves. At high prices the total demand in market *T* is limited while domestic supply is substantial, so the quantity the producer from *S* can sell in country *T* is restricted. As prices fall overall demand grows while domestic supply diminishes, so the quantity the producer from *S* can sell in country *T* increases. Thus if we abstract from differences of detail the two sale curves are of such a nature that as prices fall the saleable quantity increases.

The precise trajectory of each of these two curves, i.e. their elasticity – a sign of the expandability of sales with the fall in the price – depends on the elasticity of the two demand curves and the supply curve in *T*. The more elastic the demand curves for *A* and *B* the more rapidly the quantity of each good demanded increases with the fall in the price. On the other hand the more elastic the domestic supply, the more rapidly this supply diminishes with the fall in the price. Hence the sale of the two goods *A* and *B* extends the more rapidly for each successive fall in prices, i.e. the sale curves of these two goods are the more elastic the greater the elasticity of the two demand curves and the curve of global supply in *T*.[27]

On this premiss it is easy to resolve the point at issue here. The country *S* producer supplies the two goods *A* and *B* in country *T* at gradually decreasing prices and progressively augments sales of them. The return on the sale of each good is derived from the product of the quantity sold and its respective price, and as we have seen this return increases for a while as the price falls and then when it reaches a certain point gradually begins to fall. The sum of partial returns on *A* and *B* constitutes what I have called the supplementary global return derived from export production, and this global return too, like the partial returns from which it is derived, increases up to a given limit after which it decreases. The high point is reached in correspondence with the sale of a given quantity of good *A* and good *B*, but as production and sales increase the producer in country *S* incurs a gradually increasing global cost. Thus while the supplementary global return increases up to a given limit after which it then begins progressively to decrease, the supplementary global cost grows continuously. But the difference between the return and the supplementary cost yields the supplementary net profit on export production, so this net profit increases for a while with each rise in the quantities of *A* and *B* and then after a certain point begins to fall. Now obviously the producer in country *S* will tend to maximise this supplementary net profit, since in doing so he will maximise his total net profit, and the solution to the problem under discussion is to be found here. The maximum supplementary net profit corresponds to a given sale of *A* and *B* to which a given price of these two goods is the counterpart. Thus in offering the two goods *A* and *B* in *T* the producer in country *S* will fix precisely those prices to which the sales corresponding to the maximum supplementary net profit are the counterpart, and will adapt the level of his production to these sales. Hence whenever favourable conditions are in place for production for export, it rises to the level at which the resulting quantities available for export are such that they provide the maximum supplementary net profit in relation to the respective sale curves.

The prices that the country *S* producer fixes for sales in *T* become the prices current in the latter country. At these new prices production of *A* and *B* in *T* cannot remain as extensive as it was initially, that is before exports from country *S* began, but will have to contract, and it will contract down to the level at which the new global marginal cost in *T* becomes equal to the global price arising from the new prices of *A* and *B*. Therefore the regulatory principle of the level of export production in country *S* also determines levels of production in country *T*, and the level to which production is obliged to contract in country *S* depends on the conditions determining the maximum profit for the producer in country *T*.

But the above regulatory principle of export production gives rise to an anomaly that needs to be emphasised. We have seen that the maximum supplementary profit is reached in correspondence with the sale of determinate quantities of the two goods *A* and *B* in *T*, with these quantities being in a fixed relation to each other. To sell these quantities the producer in country *S* must produce them, and as we have seen he extends supplementary global production up to the point where he obtains the quantity of *A* and *B* indicated from the organisation of that production. But being joint, production of these two goods takes place in *S* in a constant ratio, so that production of a determinate quantity of *A* is always matched by production of a proportionately constant quantity of *B*, and under these circumstances two things may happen. The ratio in which the two goods *A* and *B* are produced may be exactly equal to the ratio in which they are sold in *T* in correspondence with the level of maximum supplementary profit, in which case the whole supplementary production both of *A* and of *B* is sold in its entirety in *T* and no complication arises. But the second possible case is that the two ratios may be different, in which case the quantity of *A* or – depending on the circumstances – the quantity of *B* that is produced is greater than the quantity that can be sold in *T* on the basis of the principle of maximum profit. Here then production for export results in an unsold residue of *A* or *B*, and this residue the producer from *S* has to destroy unless he has other markets open to him. Now of course of these two cases the one that is more likely to occur is the second, for it is only exceptionally and by pure combination that the two ratios do actually coincide. Normally they are different and thereby they create the anomaly indicated. Hence we can conclude that under the stated conditions production of two joint goods for export is accompanied as a general rule by the formation of an unsold residue of one or other of the two goods.

**12.** In the course of our study of the preceding case we saw that it occurs every time appropriate conditions are in place for production for export and that if such conditions are not present there can be no production for export. This does not however mean that export immediately becomes impossible, for there are cases in which it is equally possible, and these are the cases we must now consider.

That export can take place even without any special increase in production was noted in passing in Chapter 2 above. There we saw that given two goods produced jointly in a closed market in a monopoly regime it is exceptional for the entire production of both goods to be sold. In most cases only one of the two is sold out and there usually remains an unsold residue of the other. Now whenever such a residue is formed and, according to our hypothesis, country *S* ceases to be isolated and enters into communication with *T*, the monopolist has an opportunity to export it. Hence (and this is an anomaly of supply at joint cost) export becomes possible in this case even if production for export is not profitable, i.e. even without a special increase in production. In this case however export remains limited to just one of the two joint goods.[28]

Let *A* be the good of which a residue remains in country *S* and let it be exported to *T*. In exporting this residue the producer in country *S* is subject as we know to the sale curve of good *A* in *T* which determines the price obtainable in correspondence with any quantity exported. The producer may export the whole of the residue or only a part of it, and as a monopolist he will of course export the quantity that earns him the maximum total profit.

The total profit to the monopolist is greater to the extent that his supplementary profit from the sale of the residue is greater, but in this case, unlike in the preceding one, any element of cost lies outside the calculation of the supplementary profit. In fact the cost of the entire quantity produced and therefore of the residue is included in the computation of the maximum profit from the sale of the two goods in country *S*. Hence the supplementary profit is represented by the entire supplementary return on the sale of the residue and the monopolistic producer will tend to maximise this supplementary yield. If the maximum supplementary yield turns out to correspond to the sale of the entire residue of *A*, then the whole residue will be sold in *T*. If the maximum supplementary yield corresponds to the sale of a smaller quantity of the residue, then the residue will not all be sold, and the part of the residue that cannot be sold in *T* will in the absence of any other market be destroyed. Everything depends on the size of the initial residue and the elasticity of the sale curve. The smaller the size of the residue the greater the probability that it will be sold in its entirety and vice versa. The whole residue will be sold if in correspondence with the quantity of the residue the elasticity of the curve of sale is less than one, but if this elasticity is greater than one a part of the residue will remain unsold.

**13.** But alongside these two extreme cases, in one of which export takes place solely on account of the contribution of a supplementary production, and in the other of which it happens without any supplementary production at all, a third situation may arise from a combination and fusion of the two extreme cases, and this too has to be taken into consideration in a general theory. In those cases in which production for export is profitable there may at the same time be a residue of one of the two joint goods, for example good *A*, from domestic sales, and in this case again export takes place thanks to the combination of two different flows. Exports of good *B* involve only the quantity of *B* produced for this purpose, while exports of good *A* involve the quantity of this good produced for the purpose *plus* the initial residue, so that we have more substantial exports than in the preceding cases. It is unnecessary to add that in this case too a residue may be left from sales in country *T*, a residue either of good *A* or of good *B* depending on the conditions.

From all this it seems clear that as a result of the complications arising from supply at joint cost the problem that concerns us here presents under many different forms that give rise to a variety of solutions. If we summarise our discussion in the present chapter and list all these possible cases together with their respective solutions, we arrive at a general theory of export with discrimination of prices which may be presented in schematic form. This theory can be stated in the following four principles.

(*I*) If there is no residue in the monopolist's domestic market and if favourable conditions for production for export are absent, export even of just one of two joint goods is impossible.
(*II*) If there is a residue of one of the two joint goods on the domestic market but favourable conditions for production for export are absent, there will be export solely of the residue and of this a part may in some cases remain unsold.
(*III*) If there is no residue on the domestic market but favourable conditions for production for export are present there will be export of both goods in proportion to the quantity of each of them produced for the purpose; but the quantity of both goods produced will not always be sold in its entirety. There may in many cases be an unsold residue of one of the goods.
(*IV*) If there is an unsold residue of one of the two goods on the domestic market and favourable conditions for production for export are in place, there will be exports of both goods, and this will consist of the supplementary production *plus* the residue in question. But the entire quantity of both goods produced will not be put on the market in every case, and here too a part of one of the joint goods is quite likely to remain unsold. This demonstrates that both under a regime of linked markets and under a regime of closed markets the cases in which the entire quantity produced of two joint goods is sold are a rare exception rather than the general rule.

**Case II**

**14.** The conditions assumed in this second case are the following: monopoly of the production and sale of good *A* on the part of countries *S* and *T*, each within the limits of its own market, and competition in the two markets between the two monopolists in the sale of the other joint good, good *B*. A practical example of this case is to be met with on a daily basis in the production of gas and coke. Not normally suited to transfer over any great distance, gas is a good that is mostly consumed at the point of production. If in two countries production of gas is in the hands of two monopolists these two countries will constitute closed markets in respect of gas, but at the same time these monopolists will be producing coke – a transferrable good – and competing to sell it.[29] In this case the question is what are the laws governing the establishment of the prices of the two goods, and what are the levels to which their production will be pushed in the two countries.

To find a solution to this problem it will again be helpful to suppose that initially the two markets are completely closed and that they abruptly become linked (within limits compatible with the conditions supposed). As long as the two markets *S* and *T* are closed, production of the two goods *A* and *B* takes place in each of them under a regime of complete monopoly while production and price equilibrium is established in each independently, according to the principle discussed earlier of the maximum net profit to the monopolist. As soon as the two markets are linked the two producers lose their complete monopoly – while it continues in the case of the sale of good *A* competition enters the sphere of sales of good *B*. The question is what shifts take place in production and prices of the two goods in the two countries during the transition from the initial total monopoly to partial monopoly or a partial free market.

**15.** As long as the two markets are closed prices of *A* and *B* settle in each market at levels corresponding to the maximum total net profit for the two monopolists. Since the two markets are independent prices in one are fixed independently of prices in the other and normally differ from market to market – if they coincide it is only by chance. Hence we can immediately assume the existence of initially different prices in the two markets. The initial divergences between prices of *A* are of secondary importance since these prices will remain independent, but the initial divergences between the prices of *B* are of capital importance. We shall suppose that as long as the two markets are separate the price of *B* is lower in *S* than in *T*. Under these conditions as soon as the two markets link up it will of course become profitable to export good *B* from *S* to *T* and export will soon take place, giving rise to a transfer of money in the opposite direction with all the subsequent effects we know that leads to. In the meantime however as a consequence of the exports the price of *B* will rise in *S* and fall in *T*. If we leave aside the question of transport costs, the two prices, initially different in the two countries, will tend towards the same level and will in fact coincide from now on.

But at what level will the price of *B* settle definitively in the two countries? As the two monopolists are competing to sell *B* they cannot hold back from the market a part of the available quantity of *B* to keep its price up and each will bring the entire quantity produced to the market, where it will be sold at the price at which both markets can absorb the entire available quantity. This price will be higher or lower depending on whether the quantity available is smaller or larger. At the first moment when the two markets link up the quantity available is equal overall to the sum of the two quantities that the two countries produced when separated, but later with equilibrium re-established the quantity of *B* available will be greater or smaller than that quantity depending on whether global individual and overall production grows or diminishes as a result of the opening of the two markets. Hence the problem of the level at which the price of *B* is fixed is closely connected with the problem of the shifts in global production following on the transition of the two markets from a complete monopoly regime to a partial monopoly regime, and the solution to this second problem contains within itself the solution to the first.

The behaviour of global production in the two countries following the transition from one regime to the other depends on the principle that regulates production at the two successive points and on the particular conditions under which this principle is concretised at those points. Production is initially subject to the principle of maximum profit, and then one of the two goods moves from a monopoly regime to a free market or more precisely to a duopoly regime. But the other good goes on being sold under a monopoly regime so that production remains free of the control of the cost principle while the maximum profit principle continues to apply. The transition from one regime to the other does not therefore change the regulatory principle of production. If the regulatory principle remains the same at the two successive points, after the opening of the two markets production will tend to adjust so as to provide the two producers with the maximum net profit compatible with the condition that one of the two products is sold in a free market. Depending on whether the point of maximum profit for one or the other producer or both of them shifts or not during this transition production in each country will either grow, contract or remain stationary.

Immediately after the change of regime – i.e. before any change in production can occur – the absolute amount of the net profit of the two producers changes. The country *S* producer, selling good *B* in *T* as well and at a price higher than the initial one, obtains a net profit higher than before while the country *T* producer, selling good *B* at a price lower than the initial one, obtains a net profit lower than before. The solution to the problem continues to be centred on this point – whether or not the resulting two new net profit levels represent the maximum levels possible under the new conditions.

Let us therefore consider what the opening of the two markets really amounts to for the two monopolists. After it the country *S* producer extends sales of good *B* to the foreign market so that the demand for good *B* on which he can count, initially constituted solely by demand on the domestic market, now consists both of this demand and of demand from the other market in correspondence with each price. On the other hand after the opening of the two markets the demand for *B* on which the producer in country *T* can count, initially consisting entirely of demand from this market, comes to be constituted by this demand *minus* that part of it which is now met by the producer in the other country. Ultimately therefore the effects of the opening of the two markets are an increase in the demand for *B* to the country *S* producer and a decrease in the demand for *B* to the country *T* producer, so for research purposes it can be regarded as equivalent to this increase and decrease in demand and handled appropriately.

I gave extensive consideration to the effects of an increase or decrease in demand for one of two joint goods under a monopoly in Chapter 2,[30] so to resolve this case I need only refer to the conclusions in that chapter. In it I showed that an increase in demand for a good produced jointly with another under a monopoly regime may either cause an increase or decrease in global production or else it may fail to produce any change at all, and I also showed that all three solutions are just as liable to occur in the case of a decrease in demand.

Now if it is true that the opening of the two markets may be regarded as equivalent in terms of its effects to an increase in demand for *B* to the country *S* producer and a decrease in demand for *B* to the country *T* producer, this will mean that following the opening of the two markets global production in each of the two countries may increase, decrease or remain unchanged depending on the case. The crucial issue is that of the new curves of returns on *B*. But the curves of partial returns on *B* for the two producers are as we know only a derivation of the corresponding demand curves,[31] which as we have just observed change completely during the transition from one regime to the other. Therefore the two corresponding curves of returns on *B* also change – i.e. become distorted – and the type of distortion of these two curves – i.e. the new form they take by contrast with the preceding form – governs the effects on production. Depending on whether the two new curves are more or less steep than before, and depending on whether the point of maximum return on each of these new curves is reached in correspondence with the sale of a greater or smaller quantity of *B* than with the preceding curves, global production in each of the two countries will increase or decrease or remain unchanged. For the sake of brevity I shall not dwell on the whole range of cases where results correspond to the various conditions and to all possible combinations of these conditions – the reader is referred to Chapter 2 for that. Here I shall simply indicate for each country the probable shifts in prices corresponding to each of the three alternatives mentioned.

If in country *S* the opening of the two markets is followed by an increase in global production, this will or will not be accompanied by an increase in the quantity of *A* put on sale and therefore by a fall in the price of *A*, depending on whether the initial equilibrium at the opening of the two markets was arrived at with or without a residue of *A*. In this case in country *S* the rise in the price of *B* is matched or rather may be matched by the fall in the price of *A*.

If, still in country *S*, the opening of the two markets is followed by a decrease in global production, this will or will not be accompanied by a decrease in the quantity of *A* put on sale in *S* and therefore by a rise in the price of *A*, depending on whether the initial equilibrium at the opening of the two markets was arrived at with or without a residue of *A*, and whether the decrease in production of *A* is greater or smaller than the preceding residue if there was one. In this case in country *S* the rise in the price of *B* may be matched by a rise in the price of *A*.

Finally if in country *S* the opening of the two markets is not followed by any change in global production, the price of *A* will remain unchanged and the rise in the price of *B* will be matched by immobility in the price of *A*. Similar shifts in the price of *A* take place in country *T* in correspondence with each of the three alternatives considered above. But since in country *T* the price of *B* falls after the opening of the two markets the fall in the price of *B* will be matched by a fall or rise or absence of movement in the price of *A* in correspondence with these three alternatives. It is pointless to add that in those cases in which the opening of the two markets is accompanied by an increase in global production in country *S* or country *T* or both countries together, this increase will probably be followed by the formation of an unsold residue of *A* in each of the countries in which it takes place.[32]

Up to now I have been studying the possible alternatives separately for each country. But since after the opening of the two markets a different one of the three alternatives occurs at the same time in each country, the overall result differs from case to case depending on the different way in which these alternatives combine. The various possible cases are given by the different possible combinations of these three alternatives combined in twos. Hence after the opening of the two markets any one of the following new combinations may ensue: that global production

(1) remains unchanged in *S* but increases in *T*;
(2) remains unchanged in *S* but decreases in *T*;
(3) remains unchanged in the two countries;
(4) increases in *S* but remains unchanged in *T*;
(5) increases in *S* but decreases in *T*;
(6) increases in *S* and in *T*;
(7) decreases in *S* but increases in *T*;
(8) decreases in *S* and in *T*;
(9) decreases in *S* but remains unchanged in *T*.

As we already know, correlative price movements will take place in each of the two countries in correspondence with each of these different combinations.

In any case whatever the change in production and prices that occurs in each of the two countries, it inevitably has the effect of moving the market towards the equilibrium position corresponding to the new conditions. The resulting adjustment process is quite complicated. We have seen that from case to case the results depend on the distortion to which the curves of partial returns on *B* are subjected in each of the two countries during the transition from one regime to the other. We have also seen that these two curves derive from the respective curves of demand for *B*. Now the two curves of demand for *B* on which the two producers can count are far from having a fixed and well defined form but are in fact essentially changeable, for their shape changes as the adjustment process develops.

As global production in country *S* grows or contracts to adjust to the new conditions production of *B* in that country also grows or contracts and so do exports of good *B* to country *T*. But as imports of *B* into *T* grow or contract, the quantity of this good that the producer in *T* can sell contracts or grows, i.e. demand for *B* to this producer decreases or increases. Thus for the country *T* producer demand for *B* varies not just as a function of the market price but also in relation to changes in global production that take place in country *S* in the meantime. Likewise as global production in country *T* grows or decreases with the growth or contraction of production of *B* demand for the said good to the country *S* producer contracts or grows. Hence for this producer too demand for *B* varies not just as a function of the market price but also in relation to changes in global production taking place at the same time in country *T*. So the demand for good *B* to each of the two producers is linked to the totality of the other conditions in such a way that a change in global production in a given direction in one country causes a change in the opposite direction in demand for good *B* to the producer in the other country. Therefore as global production increases or decreases in one country the curve of demand for good *B* to the producer in the other country is distorted in consequence. Since these changes in production take place as a result of the tendency of production to adjust to new equilibrium conditions, the curve of demand for *B* to each of the two producers depends in its successive distortions on the way in which adjustment of production occurs in the other country, i.e. on the conditions that combine to determine the maximum profit in the other country. Now if it is true that the curve of returns on *B* peculiar to each producer derives from the respective curves of demand for *B*, if it is true that the form of these curves of demand gradually changes as the adjustment process goes on, if it is also true that the curve of demand for *B* depends for each producer (among other things) on the conditions that determine the maximum profit in the other country, this means that for each producer the curve of partial returns on *B* depends (among other things) on the conditions that determine the maximum profit for the other producer. But as we know the conditions of maximum profit for him depend on the distortion of the curve of partial returns on *B*. Thus the conditions that determine the maximum profit for a producer depend not just on the prevailing conditions of production, but on the conditions that determine the maximum profit for the other producer and vice versa. The two maxima therefore are interconnected and the link between them is fundamental to the general adjustment process. In fact as production by one of the two producers increases or contracts to adapt to new conditions, the point of maximum profit shifts for the other. But as the point of maximum profit for this producer shifts and the dimensions of his production change along with it, the point of maximum profit for the other shifts in response. So a mutual relation is established between the two points of maximum profit such that each shift in the one causes a correlative shift in the other. Therefore immediately after the opening of the two markets, when in the two countries the process of adjustment of production to the changed conditions gets under way, there begins this action and reaction process which ensures that as production in each of the two countries moves towards its own point of maximum profit it shifts the point of maximum profit for the other. This action and reaction process continues until production in the two countries has adjusted in such a way as to correspond to the definitive points of maximum profit for the two producers, of course within the limits where these two points of maximum profit are compatible with each other and with the other conditions of the problem.

# 4 Conclusion

Before concluding this study I shall summarise briefly the most important results of my research:

## I FREE COMPETITION IN A CLOSED MARKET

(1) Given two or more goods produced jointly, equilibrium is established at prices for individual goods such that they determine a unitary global price equal to the global marginal unitary cost of production, and such that in relation to the laws of demand for each good they cause a level of demand for each which is in the same relation to demand for the other good or goods as holds between the quantities of the various jointly produced goods (Mill's theorem).
(2) A change in demand for one or two or more joint goods that causes a shift in its price in a given direction produces a shift in the opposite direction on the part of the price of the other good or goods; and the overall shift in the various prices, if all other conditions remain the same, is smaller the greater (a) the elasticity of demand for the goods for which demand has changed; (b) the elasticity of demand for the other goods; (c) the closeness with which costs conform to the law of growing costs; (d) the number of goods produced jointly.
(3) A tax on production of one or two or more joint goods at such-and-such a rate per unit of production of the said good brings about a simultaneous rise in the price of all the joint goods i.e. it affects those exempt from it just as much as those on which it is imposed.
(4) A tax imposed on the quota system, i.e. fixed in its total amount, may be applied to one or other of the various joint goods or distributed over all at once without its effects being altered in any way. Whatever the way the tax is imposed its effect on the prices of the various goods, assuming parity in the quota, is identical in all cases.
(5) A premium on the production of a good produced jointly with others causes not just a fall in the price of the good carrying the premium but also a fall in the price of all the other goods. Assuming parity in the overall amount the way in which the premium operates makes no difference – whether it is based on global production or on

the production of one product or another the effect on prices is identical in all cases.

## II A MONOPOLY REGIME IN A CLOSED MARKET

(6) Given two or more goods produced jointly under a monopoly, equilibrium is established at those prices and in correspondence with those quantities that provide the monopolist with the maximum net profit. In these conditions however it is only in highly exceptional circumstances that the whole quantity of all the goods produced is sold. A part of the production of one or some of the goods normally remains unsold and in the absence of other markets it is destroyed.
(7) Given an initial equilibrium position the increase in demand for a good produced jointly with others raises the price of this good and there ensues a level of global production greater than, smaller than, or equal to the initial one depending on whether as a result of this price rise the curve of partial returns on this good is distorted or not; and if so, whether it is distorted in one way or another. Consequently a rise in the price of one good is not always matched by a shift in the price of the others. Depending on the case and on the various combinations a rise in the price of one good is matched sometimes by a rise and sometimes by a fall – and sometimes by neither – in the price of the others. We are dealing here with very vague and indeterminate correlations between prices, differing from case to case but nonetheless usually present. Economists like Valenti were entirely wrong to regard prices of joint goods under a monopoly regime as completely independent.
(8) A tax proportional to the quantity sold of a good produced jointly with others under a monopoly regime does not have a uniform effect on the price of the other goods as it does in a free market. Sometimes it leaves those other prices unchanged and sometimes it raises them but in very different proportions depending on the case and on the various combinations that may occur.

## III FREE COMPETITION IN OPEN MARKETS

(9) Given two or more markets initially closed in which two or more joint goods are produced, if these markets suddenly develop a link equilibrium is established between them as soon as: (a) the prices of

the various products in the various markets are on a level; (b) the global prices in the various markets coincide with the respective global marginal costs of production; (c) the quantities demanded in the various markets at the prices fixed there are equal to the respective quantities produced in each market, *plus* or *minus* the respective quantities imported or exported; (d) in each market the difference in the value of the products imported or exported equals the value of the money exported or imported in settlement.

(10) An increase in demand in a country for a good produced jointly with others brings about in all countries an increase in global production, a rise in the price of the good for which demand has increased, and a fall in the price of all the others; equilibrium is established thanks to the import into the first country of a given quantity of the good for which demand has increased and the export from the first country to the others of a given quantity of money and a given quantity of the other joint goods. The extent of the disturbance and of the transfers of goods and money is in all cases smaller (provided all other conditions remain unchanged) than that which takes place where there are independent costs. In addition in given cases and under given conditions equilibrium is established through the intervention of certain disturbing or compensatory elements that are non-existent in the case of independent costs, but the action of these elements is contrived in such a way that the extent of the overall disturbance is not altered any further. So we have a second element of the influence of supply at joint cost – it modifies certain aspects of the process by which equilibrium is reestablished in the case of independent costs.

(11) An import duty introduced in a country in favour of a good produced jointly with others functions in the same way as in the case of independent costs – namely as a protective duty stimulating production of the protected good. But as it also boosts production of the other goods at the same time it brings their price down and causes them to be exported. Hence while functioning as a protective duty on the good for whose benefit it has been imposed such a duty also functions as a premium on the export of the goods produced jointly with that one, but it exercises this function at the expense of its protective role; and the more it extends and intensifies the first function the more it lessens the effectiveness of the second.

(12) A premium on the export of a good produced jointly with others encourages both the production of it and the lowering of the prices of the others together with their export, and it resolves into a

premium on the export of all the goods produced jointly with the first. But the more it encourages the export of the other goods, the more it attenuates the efficacy of its action on behalf of the good actually carrying the premium.

## IV A MONOPOLY REGIME IN OPEN MARKETS

(13) If of two or more markets one is closed to products from the others while they are open to products from it and if in the first market production of two joint goods takes place under a monopoly regime, the basic principle according to which the monopolist regulates sales domestically and abroad is the maximum net profit principle. Exporting generally takes place at lower prices than the domestic ones and may take place either through an apposite supplementary increase in production or without it. In the first case production for export takes place only to the extent that the monopolist anticipates that it will give him an increase in his own profits (which happens under determinate conditions), and it extends up to the point at which net profit reaches the maximum level allowed by the other conditions. In the second case export takes place whenever domestic production and sales leave an unsold residue of one or the other product even though conditions favourable for production for export are not in place. In this case export takes place only of the good of which a residue remains and it is limited to that residue itself. Where we are dealing with joint goods it often happens however in both cases that the quantity of both the joint goods available cannot always be sold in its entirety even abroad so that part of the export stock remains unsold too.

# Appendices

## APPENDIX I (See Chapter 2, Part I, S. 2, Case (d))

Let $A$ and $B$ be two jointly produced goods, let $f(x)$ be the demand for $A$ and $\phi(y)$ the demand for $B$, $F(x+y)$ the global marginal unitary cost, let $k$ be the relation between the two quantities $x$ and $y$ of $A$ and $B$, and let us suppose these quantities to be constant.

$xf(x)$ will be the return on the sale of $A$, $y\phi(y)$ the return on the sale of $B$, $(x+y)\,F(x+y)$ the total cost. Equilibrium will be given by the two equations:

$$xf(x) + y\phi(y) = (x+y)F(x+y) \tag{1}$$

$$\frac{x}{y} = k \tag{2}$$

Let us now suppose that a tax is introduced that is to yield a constant fiscal product $H$. As is pointed out in the text a tax like this may be imposed in various ways: [1] on global production; [2] on production of $A$; [3] on production of $B$; [4] on production of $A$ and $B$ together. In case [1] the tax is proportional to the global product $(x+y)$, in case [2] to quantity $x$, in case [3] to quantity $y$, and in case [4] to $x$ and $y$ together.

But since the assumption is that the overall amount of the tax is always going to be $H$, the unitary tax must be different in the four different modes of application. Let $t$ be the unitary tax in case [1]; $u$ in case [2]; $v$ in case [3]; and $w$ and $z$ in case [4]. In the different cases the total amount of the tax will be:

$$(x+y)\cdot t = u\cdot x = v\cdot y = w\cdot x + z\cdot y = H$$

To ascertain the effects of the introduction of the tax and establish the new equilibrium position we need only add the tax to the other data of the problem and modify (1) in relation to the various ways in which the tax may be applied in the different cases. This will give us the following equations: in the case of a tax applied and made proportional to global production it will be:

$$xf(x) + y\phi(y) = (x+y)F(x+y) + [(x+y)t] \tag{3}$$

in the case of a tax applied and made proportional to production of $A$ it will be:

$$[xf(x) - u\cdot x] + y\phi(y) = (x+y)F(x+y) \tag{4}$$

in the case of a tax applied and made proportional to the production of B it will be:

$$xf(x) + [y\phi(y) - v\cdot y] = (x+y)F(x+y) \tag{5}$$

in the case of a tax applied at the same time to $A$ and $B$ it will be:

$$[xf(x) - w \cdot x] + [y\phi(y) - z \cdot y] = (x + y)F(x + y) \tag{6}$$

In these four equations the total tax is expressed in its constitutive elements, i.e. as a function of its unitary amount and of the production unit to which it is applied. But as it is always the same in its overall amount in all four cases, for simplicity's sake it can be represented with the same symbol $H$ in all four equations.

This then gives us (3), (4), (5), and (6) transformed into the following:

$$xf(x) + y\phi(y) = (x + y)F(x + y) + H \tag{7}$$

$$[xf(x) - H] + y\phi(y) = (x + y)F(x + y) \tag{8}$$

$$xf(x) + [y\phi(y) - H] = (x + y)F(x + y) \tag{9}$$

$$\left[xf(x) - \frac{H}{m}\right] + \left[y\phi(y) - \frac{H}{n}\right] = (x + y)F(x + y) \tag{10}$$

where: $m + n = H$

An examination of these four equations reveals that they are equivalent to each other and that by a simple transposition of term $H$ one is transformed into the other. This being the case these four equations present the same solution – i.e. they are solved in each case in correspondence with one single value of the variables $x$ and $y$ that are their roots. This means that however the tax is applied the new equilibrium is reached in correspondence with the same quantities $x$ and $y$ produced and in correspondence therefore with the same prices of $A$ and $B$.

## APPENDIX II (See Chapter 2, Part II, S. 4)

In the case of a monopoly, equilibrium is established as we know in correspondence with the quantity and the prices that give the maximum net profit. If we use $x_0$ and $y_0$ for the quantity of $A$ and $B$ produced respectively, $x_1$ and $y_1$ for the quantity of $A$ and $B$ put on sale, $x_2$ and $y_2$ for the respective residue of $A$ and $B, f(x_1)$ for demand for $A$, $\phi(y_1)$ for demand for B, $\psi(x_0 + y_0)$ for global unitary cost, the expression whose value must be maximised is the following:

$$x_1 f(x_1) + y_1\phi(y_1) - (x_0 + y_0)\psi(x_0 + y_0)$$

Let us note immediately that whenever there is no residue of $A$ or of $B$ $x_1 = x_0$ and $y_1 = y_0$. Whenever there is no residue of $A$ but there is of $B$ $x_1 = x_0$ and $y_0 = y_1 + y_2$. After this we can state the problem with regard to the various cases studied in the text.

We know that whenever $h = k$ equilibrium is established without any residue of $A$ or of $B$. Hence the expression to be maximised is the following:

$$x_0 f(x_0) + y_0\phi(y_0) - (x_0 + y_0)\psi(x_0 + y_0)$$

This expression may be rendered more briefly in the form $\Phi(x_0)$ or reduced to a function of a variable, insofar as once the value of $x_0$ is known, that of $y_0$ may be ascertained from the other equation in the system, i.e.

$$\frac{x_0}{y_0} = k$$

Hence on the basis of the theory of maxima and minima the system of equations to be solved is the following:

$$\frac{d\Phi}{dx_0} = 0 \qquad (1)$$

$$\frac{x_0}{y_0} = k \qquad (2)$$

As there are two equations in this system and two unknowns, the problem is solved.

If $h < k$ there will be a residue $x_2$ of $A$. The expression to be maximised is the following:

$$x_1 f(x_1) + y_0\phi(y_0) - (x_0 + y_0)\phi(x_0 + y_0) = \Phi(x_1, x_0)$$

Here too when we have ascertained the value of $x_0$ that of $y_0$ is found from the other equation. Hence again on the basis of the theory of maxima and minima the system of equations to be solved is the following:

$$\frac{\delta\Phi}{\delta x_1} = 0 \qquad (1)$$

$$\frac{\delta\Phi}{\delta x_0} = 0 \tag{2}$$

$$\frac{x_0}{y_0} = k \tag{3}$$

$$x_1 + x_2 = x_0 \tag{4}$$

In this case we have four equations and four unknowns $x_1, x_0, x_2, y_0$. Hence in this case too the problem is solved.

The problem whether if $h > k$ there is a residue $y_2$ of $B$ may be stated and solved in a similar manner.

Finally it may happen that there is at the same time a residue $x_2$ of $A$ and $y_2$ of $B$. This is a case that has not been considered in the text, since it is possible only on condition that the total global cost is falling, a condition that is highly exceptional if not impossible, but just to complete the series of all imaginable cases I shall take this one into consideration anyway. In such a case the expression to be maximised is

$$x_1 f(x_1) + y_1\phi(y_1) - (x_0 + y_0)\psi(x_0 + y_0) = \Phi(x_1, x_0, y_1)$$

and the system of equations to be solved is the following:

$$\frac{\delta\Phi}{\delta x_1} = 0 \tag{1}$$

$$\frac{\delta\Phi}{\delta x_0} = 0 \tag{2}$$

$$\frac{\delta\Phi}{\delta y_1} = 0 \tag{3}$$

$$\frac{x_0}{y_0} = k \tag{4}$$

$$x_1 + x_2 = x_0 \tag{5}$$

$$y_1 + y_2 = y_0 \tag{6}$$

In this case we have six unknowns $(x_i, y_i;\ i = 0, 1, 2)$ and six equations. Hence equilibrium is determined in this case too. This system of equations may be taken as the general system capable of solving the problem in all possible cases. If there is a residue of both $A$ and $B$ at the same time the system remains unchanged. If there is only a residue of $B$, $x_0 = x_1$ and unknowns $x_1$ and $x_2$ disappear from the system along with (1) and (5). If there is only a residue of $A$ $y_0 = y_1$ and unknowns $y_1$ and $y_2$ disappear from the system along with (3) and (6). Finally if there is no residue of $A$ nor of $B$ either the unknowns $y_1, y_2, x_1$ and $x_2$ disappear along with (1), (3), (5), and (6), and the system will be reduced to just the two initial equations.

## APPENDIX III (See Chapter 3, Part II, S.5)

If as has been the case in the text we use

| | |
|---|---|
| $x_{s_0}$ and $x_{t_0}$ | For the respective quantities of good *A* produced |
| $f_s(x_{s_0})$ and $f_t(x_{t_0})$ | For the respective utility curves |
| $y_{s_0}$ and $y_{t_0}$ | For the respective quantities of good *B* produced |
| $\phi_s(y_{s_0})$ and $\phi_t(y_{t_0})$ | For the respective utility curves |
| $F_s(x_{s_0}+y_{s_0})$ and $F_t(x_{t_0}+y_{t_0})$ | For the respective curves of the cost of production of the two joint goods |
| $M_s$ and $M_t$ | For the respective quantities of money in circulation |
| $k_s$ and $k_t$ | For the respective ratios in which the two goods are produced |

we shall be able to produce the following equations that express the equilibrium conditions for the production of the two goods *A* and *B* in the two countries *S* and *T* as long as they constitute two separate independent markets.

### Equations for Country S

$$\frac{x_{s_0}}{x_{s_0}+y_{s_0}}\,\frac{f_s(x_{s_0})M_s}{x_{s_0}\,y_{s_0}x_{s_0}f_s(x_{s_0})+y_{s_0}\phi_s(y_{s_0})}+\frac{y_{s_0}}{x_{s_0}+y_{s_0}}$$
$$\times\frac{\phi_s(y_{s_0})M_s}{x_{s_0}f_s(x_{s_0})+y_{s_0}\phi_s(y_{s_0})}-\frac{F_s(x_{s_0}+y_{s_0})M_s}{x_{s_0}f_s(x_{s_0})+y_{s_0}\phi_s(y_{s_0})}$$
$$\frac{x_{s_0}}{y_{s_0}}=k_s$$

### Equations for Country T

$$\frac{x_{t_0}}{x_{t_0}+y_{t_0}}\,\frac{f_t(x_{t_0})M_t}{x_{t_0}\,f_t(x_{t_0})+y_{t_0}\phi_t(y_{t_0})}+\frac{y_{t_0}}{x_{t_0}+y_{t_0}}$$
$$\times\frac{\phi_t(y_{t_0})M_t}{x_{t_0}f_t(x_{t_0})+y_{t_0}\phi_t(y_{t_0})}-\frac{F_t(x_{t_0}+y_{t_0})M_t}{x_{t_0}f_t(x_{t_0})+y_{t_0}\phi_t(y_{t_0})}$$
$$\frac{x_{t_0}}{y_{t_0}}=k_t$$

Let us now suppose that the two markets become linked and that the initial level of the two prices in the two countries is such that following the commencement of the commercial relations country *S* imports from *T* quantity $x_2$ of good *A* and exports to *T* quantity $y_2$ of *B plus* the amount $M_1$ of money. The result is that country $S$ will have available to it a quantity of good $A$, $x_{s_1} = x_{s_0} + x_2$ and a quantity of good $B$ $y_{s_0} = y_{s_1} - y_2$; country $T$ will have available to it quantity of good $A$, $x_{t_1} = x_{t_0} - x_2$ and quantity of good $B$ $y_{t_1} = y_{t_0} + y_2$. The international economic and monetary equilibrium position will be determined by the following system of equations:

$$\begin{aligned}&\frac{x_{s_0}}{x_{s_0}+y_{s_0}}\frac{f_s(x_{s_1})(M_s - M_1)}{x_{s_1}f_s(x_{s_1})+y_{s_1}\phi_s(y_{s_1})} + \frac{y_{s_0}}{x_{s_0}+y_{s_0}}\\ &\times\frac{\phi_s(y_{s_1})(M_s - M_1)}{x_{s_1}f_s(x_{s_1})+y_{s_1}\phi_s(y_{s_1})} - \frac{F_s(x_{s_0}+y_{s_0})(M_s - M_1)}{x_{s_1}f_s(x_{s_1})+y_{s_1}\phi_s(y_{s_1})}\end{aligned} \tag{1}$$

$$x_{s_1} = x_{s_0} + x_2 \tag{2}$$

$$y_{s_1} = y_{s_0} - y_2 \tag{3}$$

$$\frac{x_{s_0}}{y_{s_0}} = k_s \tag{4}$$

$$\begin{aligned}&\frac{x_{t_0}}{x_{t_0}+y_{t_0}}\frac{f_t(x_{t_1})(M_t + M_1)}{x_{t_1}f_t(x_{t_1})+y_{t_1}\phi_t(y_{t_1})} + \frac{y_{t_0}}{x_{t_0}+y_{t_0}}\times\\ &\times\frac{\phi_t(y_{t_1})(M_t + M_1)}{x_{t_1}f_t(x_{t_1})+y_{t_1}\phi_t(y_{t_1})} - \frac{F_t(x_{t_0}+y^{t_0})(M_t + M_1)}{x_{t_1}f_t(x_{t_1})+y_{t_1}\phi_t(y_{t_1})}\end{aligned} \tag{5}$$

$$x_{t_1} = x_{t_0} - x_2 \tag{6}$$

$$y_{t_1} = y_{t_0} + y_2 \tag{7}$$

$$\frac{x_{t_0}}{y_{t_0}} = k_t \tag{8}$$

$$\frac{f_s(x_{s_1})(M_s - M_1)}{x_{s_1}f_s(x_{s_1})+y_{s_1}\phi_s(y_{s_1})} - \frac{f_t(x_{t_1})(M_t + M_1)}{x_{t_1}f_t(x_{t_1})+y_{t_1}\phi_t(y_{t_1})} \tag{9}$$

$$\frac{\phi_s(y_{s_1})(M_s - M_1)}{x_{s_1}f_s(x_{s_1})+y_{s_1}\phi_s(y_{s_1})} - \frac{\phi_t(y_{t_1})(M_t + M_1)}{x_{t_1}f_t(x_{t_1})+y_{t_1}\phi_t(y_{t_1})} \tag{10}$$

$$x_2\frac{f_s(x_{s_1})(M_s - M_t)}{x_{s_1}f_s(x_{s_1})+y_{s_1}\phi_s(y_{s_1})} - y_2\frac{\phi_t(y_{t_1})(M_t + M_1)}{x_{t_1}f_t(x_{t_1})+y_{t_1}\phi_t(y_{t_1})} + M_1 \tag{11}$$

The unknowns are $x_{s_1}, x_{s_0}, x_2, x_{t_1}, x_{t_0}, y_{s_1}, y_{s_0}, y_2, y_{t_0}, y_{t_1}, M_1$ – i.e. they are eleven in number, but there are eleven equations, so the system is solvable, and equilibrium is fully determined. The eleven equations making up the system express in symbols the necessary conditions for international economic and monetary equilibrium to be established. These conditions are as follows:

[1] In each country after exchange of goods and money has taken place between the two countries the new respective global price must be equal to the new respective marginal cost (1), (5).
[2] In the two respective countries the quantities of *A* and *B* respectively produced must be equal to the respective quantities sold (or consumed) *plus* or *minus* the quantities imported or exported (2), (6), (3), (7).
[3] There must be a constant ratio between the quantities of *A* and *B* respectively produced in the two countries (4), (8).
[4] After there has been exchange, prices of *A* and *B* in the two countries must be equal (9), (10). On the basis of the conditions set out above of an equality between the new global price and the new marginal cost in each country this equality between the prices gives rise to the other condition.
[5] The new global marginal costs in the two countries must be equal.
[6] The value of imports of *A* into *S* and the value of exports of *B* *plus* the circulating medium $M_1$ exported from *A* must be equal (11).

# Notes and References

**Foreword by Lionello F. Punzo**

1. Schumpeter (1954), p. 855.
2. Schumpeter (1954), p. 855. The single most important representative of this line of thought, joining theory with applied analysis, is obviously L. Einaudi. Actually, the Italian landscape was more rugged than Schumpeter says, as Bagiotti observes:

   > Fanno's Italian contemporaries, with various degrees of subtelty and individuality, stood between the tendencies of historical or pure economics (with a few lags and revivals), and classical Marxist criticism. Pure economics was processed in terms of both causalism and functionalism, with various functionalistic factions forming around the personality and the writings of Pareto. [And specifically] Fanno, whose rapid career had been sponsored by Loria, reflects some of his tastes at his beginnings. But other inspirations, other themes, and the personality of Pantaleoni, versatile and extemporaneous, constrained him to turn to profounder theoretical considerations. (All citations from Bagiotti, 1966a, pp. xxi–xxii)

   For more details on the history of Italian thought see Finoia (1980).
3. As pointed out by Legris (1994).
4. Bagiotti (1966a). For Fanno's model of fluctuation, see the contribution by Montesano who associates it with the exogenistic model of Frisch, while Cozzi gives an endogenistic interpretation, both in Bagiotti (1966a). Halevi (1987) accepts Montesano's interpretation.
5. See Bagiotti (1966a) for a description of Fanno's works, Shoup's chapter is devoted to Fanno's theory of Joint Cost. See also Bartoli (1981) and Manfredini Gasparetto (1992).
6. These are: 'Correlazioni tra prezzi e curve statistiche di domanda e offerta' (1932), and 'Della identità di alcuni teoremi di economia pura' (1934).
7. Fanno in his obituary for Marshall (Fanno, 1925, p. 173). On the influence of Marshall, see the illuminating words in Legris (1994).
8. '[This is] a classification which does not reflect real objective differences, but like all classifications it follows the kind of clear and precise qualitative dividing lines that simply do not exist as such in the real world., (Fanno, 1914), p. 12.
9. I agree with Bagiotti's remark (1966a, p. xxii) that Schumpeter has overstressed the link between joint supply and general equilibrium.
10. Marshall's *Principles* (1907), Appendix to Bk V.
11. From the standpoint and within the precincts of partial analysis, this universal interdependence was demonstrated by numerous investigations

into the theory of joint and composite demand and supply and of the values of related commodities which produced some of the most illuminating passages of Book V of the *Principles* and which further developed by Edgeworth. In fact, it stands to reason that the comprehensive but gaunt and *colorless* idea of the universal interdependence... – and so easily provokes sneers about everything depending upon everything else – can be brought home and made alive to many by means of concrete cases about the relations between the values of beef and mutton or again of tea and sugar...And this can be done by violating the restrictions inherent in the method of partial analysis. (Schumpeter, 1954, pp. 995–6).

12. In Italy, where the adherents to the theory of economic equilibrium began to form faction, the gist of the theory of these monographs had to seem somewhat spurious (Bagiotti, 1966a, p. xxii).
13. Schumpeter (1954), p. 996, n. 12.
14. *QJE* (1913), p. 537.
15. *QJE* (1913), p. 691.
16. *QJE* (1913), p. 538.
17. In Taussig's definition 'supply at joint cost includes supply of goods or services that can take place only connectedly and supply of goods or services which could be produced separately' (Fanno, 1926, p. 22).
18. Fanno (1926), pp. 21–2.
19. *QJE* (1914), p. 684.
20. The problem originates much earlier, and found a treatment in J. Rae (1834) and later in Akermann's much celebrated doctoral dissertation, reviewed by Wicksell in an appendix of his *Lectures on Political Economy* (1934).
21. As acknowledged by Edgeworth in his comments to Fanno's work, in Edgeworth (1915).
22. As suggested by Jevons, Edgeworth himself, and more recently by the neo-Ricardians, e.g. Steedman (1984), Kurz (1986), Schefold (1987), to name only some of them. Moreover, according to Kurz, results holding for the case of non-joint production are not relevant for they cannot be carried over to this 'more general case'.
23. This position could be associated with the name of Marshall (Book V), who suggests that 'it is hard to find examples of [joint production] where the quantities of individual commodities could not be changed at the margin without affecting the quantity of the jointly produced good'.
24. Sraffa (1960), p. 63.
25. Fanno (1914), p. 7.
26. Fanno (1914), p. 41.
27. Obviously, this is an *ex post* rationalisation, for none of the economists involved had the mathematical apparatus nor skill to handle the problem as it should have been handled. Actually, the appropriate mathematical tools, which eventually found employment for the purpose (with von Neumann, A. Wald and others), had just been created. Brouwer's fixed-point theorem was published in 1912!
28. Actually, the introduction of demand would make the situation worse!

29. Fanno (1914), p. 39.
30. *Palgrave's Dictionary* (Mill, 1910b), p. 485.
31. This may or may not happen, and on purely mathematical grounds, it can be argued that the matrix is likely to be rectangular. This is because there is no *a priori* relation between rows and columns. For a different opinion, see Bidard (1986).
32. Obviously, a system determined in the algebraic sense does not necessarily yield economically – i.e. non-negative solutions in prices – and further assumptions, on the viability of the technology, will have to be introduced.
33. Goodwin and Punzo (1987); Punzo and Velupillai (1984).
34. On the literature, see Salvadori and Steedman (1988); Schefold (1980, 1987).
35. There is no reason to exclude this case, as is implicitly done in the discussion referred to in n. 34. This is, by the way, the case taken up by Sraffa.
36. A peculiar assumption would be required to exclude the possibility of free goods. Relative quantities demanded should be exactly equal to relative quantities produced – i.e. there is linear dependence between net output and demand vectors.

**Foreword by Michio Morishima**

1. More accurately, Hicks himself never specified the pattern of substitution and complementarity relationships so as to satisfy the chain rules. In spite of this I ascribe the origin of the rules to him.
2. We may mention as an example of the *M* system an economy of tennis racket, squash rackets, tennis balls and squash balls being traded, and as an example of the anti-*M* system an economy of tennis rackets, tennis balls, golf clubs and golfballs being on the market.
3. Becker (1965, pp. 493–517) used a similar model in discussing the household's allocation of time.
4. In this volume Fanno makes a parallel, international economic analysis under the assumption of perfect competition. In the same manner as I have in the previous sections replaced his assumption that competitors work in different two countries by the one in which they both work in the same country, I will continue to make the same assumption of a closed economy for analysing joint production under monopoly, in order to eliminate the problems concerning the complexity of foreign exchange ratios.
5. Another kind of joint production dealt with von Neumann (1945–6) and Sraffa (1960) is, of course, concerned with new and aged capital goods. In regard to this problem too, Sraffa neglects the demand–supply equilibrium condition (10). Besides, he insists that the idea of treating the fixed capital goods left over for production in the future as a kind of joint product was 'introduced' by Robert Torrens and then was 'adopted' by Ricardo, Malthus and Marx. (see Sraffa, 1963, pp. 94–5). We can nevertheless, find no trace of their efforts to determine the

value of the fixed capital goods left over for the next period by using the proceed–cost conditions (11) with the output matrix $B$ allowing for joint outputs. We find rather that Torrens gets that value on the basis of the assumed rates of depreciation. Sraffa's proposition, though it is supported and respected by most Sraffians, has no documentary backing (see Morishima, 1996, section 4).

**Introduction**

1. Cournot (1887).
2. Jevons (1931).
3. Edgeworth (1881).
4. Marshall (1907).
5. Auspitz and Lieben (1889).
6. Pantaleoni (1957), p. 164.
7. Mill (1909), Bk III, ch. 16, p. 569.
8. 'Thus then, strictly speaking there are only connected prices and the theory of connected prices, which usually looks like an unusual and complicated case, is in fact the only complete theory of prices ... Every good has its own curve of demand and its own curve of supply, and both are variable at any given moment, if only by virtue of the connections between variations in demand for and supply of any one good and variations in demand for and supply of all the others.' Pantaleoni (1909), pp. 172, 250.
9. Jevons (1931), p. 198.
10. Fisher (1892).
11. Walras (1954); Pareto (1896–7, 1971).
12. Pareto (1892, p. 205; 1893, p. 307; 1896–7, ii, p 74); Barone (1894a), pp. 414, 435; Anpetit (1901), p. 44.
13. Barone (1894b, p. 215; 1894c, p. 408); Ricci (1906), p. 230; Fisher (1912), pp. 287 ff.
14. Pareto (1971), p. 112.
15. Pantaleoni (1957), pp. 81–2.
16. Ferrara (1856), p. lxix.
17. See Fanno (1913), esp. Part II.
18. Fanno (1913), pp. 283–4.
19. Pareto (1971), p. 112.
20. Strictly speaking this result is not inevitable. It may in fact happen that individual price shifts occurring on the basis of initial prices are such that, taking account of the quantities of the various goods, they balance each other out perfectly. In this case the mass of circulating goods keeps its initial overall value. But a case like this is so unlikely to happen that it can be ignored. Hence as a general rule it may be stated with confidence that in the first phase of the adjustment process every disturbance in the level of marginal utility of a good tends to change the overall value of circulating goods.
21. Research on this point will supplement the research on the laws of the money market carried out in the above work. There my concern was

with the statics and the dynamics of the general level of prices and I paid no attention to individual prices that produce that level nor to the individual shifts in those prices. In this work however I shall be concerned at the appropriate place with the movements of individual prices occurring at the time of disturbances that change the general level or that modify individual prices while keeping the general level unchanged. Hence the two studies are complementary and together help to clarify the two fundamental aspects of the laws of prices.

22. To anticipate for a moment a subject that will be fully dealt with below, let me say that in the case of disturbances generally tending to bring monetary tension into the first phase of the adjustment process, prices of goods whose level of marginal utility has risen will in the new equilibrium position increase less than proportionately to the rise in their marginal utility; prices of goods whose level of marginal utility falls will fall proportionately more than the level of marginal utility. In the case of disturbances tending to import monetary depression into the first phase of the adjustment process individual prices of goods whose level of marginal utility has risen will increase more than proportionately to the increase in the level of marginal uitlity; prices of goods whose marginal utility has fallen will go down less than proportionately to that decrese.

## 1 Preliminary Observations on Supply at Joint Cost

1. 'Two products are supplied jointly when a unit of investment expended upon increasing the normal output of one, *necessarily* increases that of the other also...' Pigou (1912, p. 215).
2. 'Whenever a very large fixed capital is used not for a single purpose, but for varied purposes, the influence of the principle of joint cost shows itself...' Taussig (1911, i, p. 221). The typical example of production at joint cost in this sense is to be found in the railway industry which by one single production process simultaneously produces the transport of goods and the transport of passengers, the transport of a wide variety of goods and the transport of passengers of different classes (see Taussig, 1911, ii, p. 369).
3. On the extension of the concept of joint cost a lively but objective polemic has recently arisen between Pigou and Taussig. Pigou is a supporter of the narrow concept and Taussig of the wide. Edgeworth accepts the one or the other depending on the case. (Edgeworth, 1925, i, p. 126; 1913, p. 218). Pigou issists that it is a feature of joint cost that production of the different goods must take place at the same time, but this Taussig (1913, p. 693) disputes: Professor Pigou finds joint costs only where it is impossible to increase the supply of one commodity without at the same time increasing the supply of another. I submit that the principle of joint cost may be applicable even though a supply of one thing does not necessarily entail the supply of another. The debate continued over several months and the articles written

by the two authors in succession on the topic appeared in the February, May and August 1903 numbers of the *Quarterly Journal of Economics*.

4. Pigou (1913a), p. 691.
5. Mill (1909), Bk III, Ch. 16, pp. 569–71.
6. See esp. Cournot (1887), pp. 99–116; Sidgwick (1887), p. 198; Nicholson (1897), ii, p. 52; Graziani (1908), p. 256; Valenti (1909), pp. 315–23; Flux (1904), pp. 64–6; Taussig (1911), 1, pp. 218–21; Fisher (1912), p. 348; Edgeworth (1910a), p. 485.
7. Marshall (1907), p. 388.
8. The study of supply at joint cost has recently attracted the attention of theorists of railway economics like Ripley, Edgeworth, Taussig and others.
9. The approximations referred to in the text here are not to be confused with the three successive approximations spoken of in the Introduction. These latter relate to the overall development of the work. The former represent the successive stages in the development of the study of supply at joint cost, which are part of the first approximation of the overall work.
10. One exception is special categories of products and particularly those luxury products that confer on whoever uses them a level of distinction over against other social classes. Demand for these goods from the upper classes varies both in relation to the direct utility such goods have for them and in relation to the quantity possessed by the other classes, and this demand is of course greater to the extent that the latter quantity is smaller. Now if all other conditions remain unchanged a good is more commonly used to the extent that its price goes down. But for determinate social classes the demand for determinate products diminishes rather than increases with a fall in the price, and in all such cases therefore we are faced with an exception to the general rule. This is an exception that does not however significantly modify the trajectory of the overall curve of market demand although it does invert the normal trajectory of some individual curves of demand. In fact in these cases under consideration, following the fall in the price the reduction in demand from certain social classes is accompanied by an increase in demand from the more numerous classes, and this means that overall the demand from the whole market also follows the tendency to increase with the fall in price. Hence even in those very special cases where some individual curves of demand present a trajectory that is the inverse of the norm, the curves of demand of the whole market follow the normal trajectory overall. On this, see Cunynghame (1892, pp. 35ff.); Pigou (1903, pp. 60ff.) and (1913b, pp. 19–24).
11. Marshall (1909), Bk III, ch. 4, pp. 102–3 and Mathematical Note 3, p. 839. See also Pigou (1912), p. 92; Edgeworth (1910b), i, p. 691. If $y = f(x)$ is the equation of a curve of demand in which the variable $x$ represents the available quantity of a good and values $y$ of the function represent the corresponding demand prices, the elasticity of demand as denoted by $i$ is given by the following expression:

$$i = \frac{f(x)}{x}\frac{dx}{dy} \text{ i.e. by } i = \frac{f(x)}{xf'(x)}$$

On this, see Pigou (1910), p. 637; Barone (1912), p. 71.

12. For the ambiguities that may arise from too much vagueness and imprecision in speaking of the elasticity of demand see Jannaccone (1902), pp. 15–21.
13. Cairnes (1874), pp. 43–109.
14. Jannaccone, P. (1901), pp. 67, 336 where the author shows how at the margin of production the social and individual cost of production, the physio–psychological cost and the monetary cost sustained by the company coincide.
15. Barone (1908), i, p. 18, or ii, p. 160.
16. Pantaleoni (1957), p. 192.
17. The newer idea of expenditure pushed up to the margin of profitableness, in a regime of competition, implies the law of increased cost. The law of diminishing costs, as Cournot argues, is only intelligible on the supposition of monopoly. (Edgeworth 1925, pp. 87–8).
    See also Marshall (1920) Bk V, ch. 12, p. 459 n.1 and Mathematical Note XIV, pp. 846 ff; Fisher (1912), p. 320; Flux (1904), pp. 59 ff.; Cunynghame (1892), p. 89.
18. Average costs remain constant as long as marginal costs remain equal to average costs, in which case of course marginal costs are constant too. On the other hand average costs keep on rising until the cost of the increased amount produced (marginal cost) is greater than the average cost without this increased amount; which may happen either if marginal costs are constant or if they too are rising or falling. And average costs keep on falling as long as the marginal cost is less than the average cost; which may also be the case either if marginal costs are constant or if they are falling or rising. (Jannaccone, 1914, p. 264)
19. Jannaccone, (1914), p. 263.
20. Pigou (1912), p. 173. On the nature of the relationship between these two curves see also Edgeworth (1913), pp. 210–11; cf. also Flux (1904), pp. 47 ff; Marshall (1909), BkV, p. 344; Cunynghame (1892), pp. 66 ff.

**2 The Theory of Supply at Joint Cost: The Case of a Closed Market**

1. I will include profits in the cost of production – according to Mill (1909, BK II, ch. 15) and his followers it includes the interest on capital employed, the premium attached to the risk and the remuneration of the entrepreneur's labour as a manager. If we consider the normal average rate of profit, given in the majority of cases the necessarily exact correspondence between premium and risk which means that the premiums calculated are equivalent to the losses suffered by the risks, the premium lies outside the component elements of normal profit. If we follow modern writers in separating the person of the capitalist from that of the entrepreneur, it is likewise necessary to separate interest on the capital from the remuneration of the labour of managing. Thus in

this case profit strictly so called is reduced to the mere remuneration of the labour of managing, a remuneration which, like that of the other coefficients of production, is commensurate with the level of marginal productivity. This principle on the basis of which in a limited market equilibrium state the entrepreneur makes neither profit nor loss and simply receives remuneration for his labour as manager, espoused by Walras (1954, pp. 225 ff.) is accepted by Pareto (1896, i, pp. 40, 57 ff.), Barone (1896, pp. 141, 145) and others. Clark, who lays down the principle of marginal productivity on the basis of the distributive process, excessively restricting the economic concept of profit, asserts that in static economics profit is zero and that it is a phenomenon exclusive to dynamic economics. (Clark, 1908, pp. 203–4; 1907, p. 87). W. Smart [1912] too, though less correctly than Clark, bases the principle of marginal productivity on the distributive process.

2. If 100 units of good $A$ and 50 units of good $B$ are obtained from global production, $\frac{100}{150}$ units of good $A$ and $\frac{50}{150}$ of good $B$ will be obtained from each unit of global production. Hence if we use $p_a$ for the price of $A$ and $p_b$ for the price of $B$ the return corresponding to each global unit of production – what we have called the *global price* – will be

$$\frac{100}{150}p_a + \frac{50}{150}p_b$$

3. To avoid misunderstanding it is necessary to clarify the meaning of the variable $x+y$. This represents the overall quantity of $A$ and $B$ to which the global unitary cost $F(x+y)$ corresponds. But the expression $x+y$ is not to be considered as constituted by two variables. Nor could it be otherwise, for $x$ and $y$ are linked in such a way that a variation in one of them involves a variation in the other and vice versa, so that the function $F(x+y)$ is to be considered as a function with one single variable. $x+y$ is therefore in fact a complex variable that expresses a given extent of global production from which is obtained quantity $x$ of $A$ and quantity $y$ of $B$. Perhaps it would have been more correct to indicate this global production by a distinctive symbol instead of by the symbols $A$ and $B$, but this would have introduced a complication. The introduction of a further unknown would have made necessary the introduction of another equation into the system, something we preferred to avoid. Hence the reader is warned that the complex variable $x+y$ is only an expedient to represent a given extent of global production. If $x+y$ is the number of units that constitutes global production then $x/(x+y)$ and $y/(x+y)$ are respectively fractions of a unit of $A$ and $B$ that are obtained from each unit of global production. Hence

$$\frac{x}{x+y}f(x) + \frac{y}{x+y}\phi(y)$$

will be the *global unitary price* obtained for each unit of global production, if this price is understood in the sense indicated in the text.

4. This conclusion, with due variants, additions and corrections is very similar to that of Mill (1909), Bk II, ch. 15, p. 387.

5. For the present I am discussing the problem of joint costs on the basis of the demand curves for jointly produced goods being independent of one another. Clearly this hypothesis does not correspond precisely to reality, but I will nonetheless assume it to be applicable for the time being. I shall not be in a position to complete my discussion of this point in a further approximation until I have discussed the correlations established between prices through their respective demand levels.
6. Auspitz and Lieben (1889), pp. 41–6.
7. Here I have in mind the case in which a change in the law of global cost leaves unchanged the proportion in which the two goods are produced, i.e. leaves unchanged the constant $k$. The case in which the change in the cost is accompanied by a change in the value of $k$ will be considered in paragraph (g) below.
8. Thus if for example demand for $A$ is more elastic than demand for $B$ and if $x < y$ the proportional shift in the price of $A$ will be decidedly smaller than the shift in the price of $B$, since the difference in elasticity of demand and the difference in the quantity produced act in the same direction, i.e. tend each separately and both cumulatively to render the shift in the price of $A$ less than the shift in the price of $B$.
9. If demand for $A$ is more elastic than demand for $B$ and if $x$ is greater than $y$ we find ourselves in the presence of two conflicting factors. Whereas by reason of greater elasticity the price of $A$ tends to shift less substantially than that of $B$, by reason of the greater quantity produced $A$ tends to shift more than $B$. Hence it is impossible to affirm as a general rule which of the two prices actually shifts the more. In some cases it will be $A$, in other cases $B$.
10. The total cost is equal to the global marginal cost multiplied by the quantity of global production. This might seem not to be correct at first sight; for unless we are talking about production at constant costs, the curve of marginal costs is a rising curve, so that the totality of expense incurred is equal to the integral defined by the function of the marginal costs. (see above Chapter 1, s. 3). But just as the global price is equal to the global marginal cost so the total return will be equal to the price for the quantity, i.e. it will be higher than the expenses incurred. The plus represents differential yields earned by producers who find themselves in more advantageous conditions. As the unitary price is equal to the marginal unitary cost, equilibrium will be established in correspondence with a total return equal to the unitary cost multiplied by the quantity. This means that if there is to be equilibrium the total return must be such as to compensate not only for the expenses incurred but also for the differential yields. If therefore these expenses do not form part of the marginal cost, nor therefore of the price (as the great Ricardo demonstrated) they do form part of the total cost, i.e. that total cost which is counterbalanced by the total return as an equilibrium factor. Hence it is correct to calculate the total cost in the manner indicated.
11. For a more rigorous demonstration of the thesis defended here see Appendix I.
12. Pantaleoni (1910), pp. 292–4.

13. Once this hypothesis is abandoned in order to bring the theory closer to reality and it is assumed that the transformation of the production process involves cost, i.e. is accompanied e.g. by the loss of a part of the capital invested and by the need to invest new capital, then this cost has to be included in the list of the conditions crucial to the possibility or otherwise of the accomplishment of the transformation and it has to be calculated in a particular manner and on the basis of particular criteria, neither of which I have the space to go into here.
14. Global production in the new production process can be represented by the expression: $[(x_0 + mx_0) + (y_0 - ny_0)]$, giving this expression the conventional meaning as expounded in chap. 2, part I, section 2. In relation to the demonstration I am putting forward here I shall suppose that $[(x_0 + mx_0) + (y_0 - ny_0)] = (x_0 + y_0)$, i.e. the comparisons made between the two production processes to decide which is the more profitable always refer to a global production that is identical in both cases.
15. The replacement of the production process will be possible whenever:

$$\frac{(x_0 + mx_0)}{(x_0 + mx_0) + (y_0 - ny_0)} f(x_0 + mx_0) + \frac{(y_0 - ny_0)}{(x_0 + mx_0) + (y_0 - ny_0)} \varphi(y_0 - ny_0) > F(x_0 + y_0)$$

16. If we use $F_1(x + y)$ to represent the new function of marginal costs, the replacement of the process will be possible whenever:

$$\frac{(x_0 + mx_0)}{(x_0 + mx_0) + (y_0 - ny_0)} f(x_0 + mx_0) + \frac{(y_0 - ny_0)}{(x_0 + mx_0) + (y_0 - ny_0)} \varphi(y_0 - ny_0) > F_1[(x_0 + mx_0) + (y_0 - ny_0)]$$

i.e. for: $[(x_0 + mx_0) + (y_0 - ny_0)] = (x_0 + y_0)$

17. If in the first production process a global production of 150 gives 100 units of $A$ and 50 of $B$, the price of $A$ is $\frac{100}{150}$ of the global price while the price of $B$ is $\frac{50}{150}$ of it.

If in the second production process in a global production again of 150 units 125 units of $A$ and 25 of $B$ are produced – the quota part of the price of $A$ will be $\frac{125}{150}$, while that of B falls to $\frac{25}{150}$.

Hence the increase or decrease in the quota parts of prices vary in relation to the absolute variations in the quantities of the individual goods.
18. Goods of the same kind but of varying quality are substitutes for each other, and as such their prices are linked by determinate relations of interdependence. So far in this study we are not really ready to study

relations of this kind (which will be considered at length elsewhere) and we cannot therefore take them into consideration as yet. As a first approximation we shall when considering goods of varying quality take account only of the relations between their prices that result from the fact that they are produced jointly.

19. The products butter and skim milk are not yielded in invariable proportion. Under these circumstances the cost of a given amount of the single products can be determined in addition to their joint cost. The richer milk can be secured at an increased cost, and the question to be considered is whether it is worth while. If it be wholly a question of whether it is worth while from the point of view of whether the increase of expenses is worth incurring, the direct comparison of the increased cost of securing the richer milk with the value of it when secured yields the answer. (Flux, 1904, p. 67)
See also Marshall (1907), p. 390.

20. See above Chapter 1, s. 3 for discussion of the way unitary cost is calculated in this case and therefore for the meaning of the symbols $g(x+y)$. But here we must point out that in this case of a monopoly, unlike in the case of free competition, we do not include profit in the unitary cost of production. Therefore the expression $g(x+y)$ refers only to unitary expenses actually incurred by the company from production. We do this to avoid complications in our calculations. If cost included the entrepreneur's profit, his total profit would consist in the sum of profits taken and the sum of the extra gain from the monopoly. These two sums ought to be computed separately in the determination of the total profit to be maximised – which would cause obvious complications in the calculation. Instead I shall exclude profit from cost and assume that total profit comes purely from the difference between cost and total returns, which makes the relevant calculations easier and plainer.

21. Edgeworth (1898), p. 236; Cunynghame (1892) pp. 121 ff.

22. Barone (1908), ii, p. 142.

23. Jannaccone (1914), p. 263.

24. Cournot thinks it would be *absurd* that the absolute expense of production should decrease as production increases. (Cournot, 1887, p. 59)

25. Taussig saw the enormous complications that production at joint cost introduces into the theory of value under a monopoly regime – but he was not concerned to study these complications; he merely indicated their existence. This is what he wrote on the topic:
Either monopoly alone or joint cost alone entails consequences for value which diverge far from the simpler cases. When the two are combined, a variety of interacting forces must be considered – joint and separate cost, marginal utility and elasticity of demand, monopoly and maximum profit and the effects upon monopoly of possible competition. (Taussig, 1911, i, p. 221)

26. By 'the incremental ratio of the total return or of partial returns or of the total cost' I mean the increase in the total return, partial returns, or total cost for every increase in global production. And this expression is to be understood in this sense in the various cases in which it is employed here.

27. The only author as far as I know who has noticed this anomaly involved in production at joint cost under a monopoly regime is Pigou. But he makes a brief passing reference to it in a note without any emphasis and without drawing all the conclusions and consequences from it that he could have done. His comment is: When two products are yielded by the same process, conditions are conceivable, under which simple competition would evolve no output, but simple monopoly, coupled with the destruction of a part of one of the products, would yield a considerable output. (Pigou, 1912, p. 196, n. 1)
28. This strange anomaly arising from production at joint cost, according to which in determinate cases the attainment of maximum monopoly profit is subordinate to the destruction of a part of one of the two goods produced, suggests certain reflections not really central to the case currently being considered of a strictly closed market, but having at least some relevance to it. The destruction of a part of the goods produced jointly is in fact only admissible in the context of a strictly closed market; it is the final option open to the monopolist after all others have been tried. And even though we have here an exception to the closed market hypothesis, it is clear that the monopolist finds it profitable to sell abroad the residue that has not been sold at home, and indeed to sell it at any price however low rather than to destroy it. Hence supply at joint cost under a monopoly regime gives rise to strange cases of export that merit study for the peculiar conditions in which they occur. I shall recur to this in Chapter 3 when I deal with the case of monopoly in relation to linked markets.
29. For a rigorous demonstration of the assertion that equilibrium is determined in all the cases see Appendix 2.
30. Auspitz and Lieben (1889), pp. 41–6.
31. Let *OAS* be the initial curve of returns on good *A* (see Figure 1). The maximum return is reached at point *A* and so in correspondence with quantity *OB*. If demand for *A* increases the new curve of returns on *A* may equally well assume either the form *OCT* or *OEU*. In the first case the quantity corresponding to the maximum return reduces to *OD*, in the second case it increases to *OF*.

    The same anomalies may also occur in the case of a reduction in demand. Let *OAS* (Figure 2) be the initial curve of the returns on good *A*. The highest return is reached at point *A*, i.e. in correspondence with quantity *OB*. If demand for *A* falls the new curves of demand for *A* may equally well assume the form *OCT* or *OEU*. In the first case the quantity corresponding to the maximum return decreases to *OD*, in the second it increases to *OF*.
32. Let *OAS* (see Figure 3) be the initial curve of returns on *A* that has the point of maximum return on *A* in correspondence with quantity *OB*. Given the trajectory of the curve of costs and of that of returns on *B*, initial equilibrium is established in correspondence with point *C* on the said curve and at the quantity *OD* of good *A*. If demand for *A* increases the new curve of returns on *A* may equally well take the form *OET* or *OFU*. If it takes the first of these forms, in correspondence with quantity *OD* the steepness of the said curve (i.e. the incremental ratio

of the return on $A$) will be less than the steepenss of the previous curve and it will be profitable to reduce production. If the new curve of returns on $A$ takes the second form, in correspondence with initial equilibrium quantity $OD$ it is steeper than the initial curve and an increase in production is profitable.

Similar results may occur in the case of a reduction in demand for $A$. Let $OAS$ (Figure 4) be the initial curve of the return on $A$ and let $C$ be the point on the said curve at which initial equilibrium is established in correspondence with quantity $OD$ of $A$. If demand for $A$ falls the new curve of returns on $A$ may equally well take on the form $OEU$ or the form $OFT$. In the former case in correspondence with quantity $OD$ the curve is steeper at point $E$ than before, and it will be profitable in spite of the fall in the demand for and the price of $A$ to increase global production. In the latter case the curve at point $F$, i.e. in correspondence with quantity $OD$, is less steep than before and it will be profitable to cut global production.

33. Thus for example Valenti:
If none of the goods produced jointly is subjected to competition, then their value will be governed by the level of quantitative limitation of each as if we were dealing with goods produced independently of each other. (Valenti, 1909, p. 325)
34. Cournot (1887) p. 68; Marshall (1907), p. 482; Barone (1908), ii, p. 143.
35. Professor Jannaccone (1902, p. 21) notes quite rightly that it is inaccurate to speak of the transposition of a tax under a monopoly regime. Rather than a genuine transposition there is a shift in prices which cannot be called transposition given the particular character of the monopolistic regime. Edgeworth (1910c, p. 300, n.1) has recently accepted Jannaccone's argument.

## 3 The Theory of Supply at Joint Cost: The Case of Linked Markets

1. Pantaleoni (1957) p. 84; Menger (1976), p. 57.
2. Only consumable goods have an elementary ophelimity that is proper to them. It is only by convention that the notion of ophelimity is extended to cover non-consumables so that the elementary ophelimity of consumables which they procure for us is attributed to them. In other words only direct goods have an elementary ophelimity that is proper to them; instrumental goods have that of the direct goods they procure (Pareto, 1896, i, p. 42).
3. However the way in which money and other instrumental goods derive their utility from the utility of direct goods is different. Instrumental goods in general need a certain time to transform themselves into direct goods. Their utility is therefore for the most part prospective – precisely because it is not exactly equal to the utility of the direct goods which the instrumental goods serve to acquire, but instead equal to such utility *minus* the depreciation that for well known psychological reasons prospective utilities suffer from compared with present utilities. But unlike other instrumental goods money is always at any moment

convertible into direct goods, and its utility is thus always at least potentially a present utility. As such it is always exactly equal to the utility of the goods it acquires, i.e. it does not undergo any depreciation in comparison with that utility. (On this see Pantaleoni (1957), p. 90; Landry (1904), p. 57; von Böhm-Bawerk (1902), ii, p. 185; von Böhm-Bawerk (1891), p. 247; Fisher (1907), p. 88.)

4. Pantaleoni (1957), p. 71; Menger (1976), p. 132.
5. In the equilibrium state the price of two goods is in each case equal to the relation of the force of the need left after the exchange of these two goods in any reasoning individual (Aupetit, 1901, p. 86)

   The marginal utilities of all articles consumed by a given individual are proportional to the marginal utilities of the same series of articles for each other consumer, and this uniform continuous ratio is the scale of prices of those articles. (Fisher, 1892, p. 37) See also Schumpeter (1908), pp. 130–1; Walras (1957), p. 107.
6. Here I am deliberately avoiding the term *total utility*, which in the scientific terminology currently predominant is used to designate something very different, namely the sum of the successive marginal utilities of a good for successively increasing quantities of that good. It is represented mathematically by the definite integral (by 0 and $x$ units of the good under consideration) of the function of the curve of marginal utility of the said good (Fisher, 1892, p. 18; Aupetit, 1901, p. 46).
7. Walras (1900) p. 331.
8. More briefly this sum could be called the *weighted sum* of the marginal utilities of the various goods, and I recommend this terminology to scholars who may well want to consider adopting it. In the text however I shall continue to designate this sum with the periphrasis used up to now, to avoid the mistakes that often arise from a new terminology that is not yet in general usage.
9. It is this method rather than the tortuous methods suggested and followed by writers like Kinley (1911) and G. del Vecchio (1909) that enables us to relate the law of the value of money to the general law of value, i.e. to the principle of the marginal level of utility. The concept of a marginal level of utility of money as expounded by these writers, resolving into the profit from marginal exchange obtained by means of money, introduces into the theory of circulation confused and incomprehensible concepts that complicate the problem without giving any satisfactory solution to it. The concept I have put forward coincides with the dominant concept of the marginal utility of instrumental goods and the results obtained by developing this concept fit in perfectly with all the various theories of the value of money.

   The principle in question as expounded in the text presents an outline of the law of the value of money in relation to the four hypothetical conditions that for simplicity's sake I have made the foundations of my research. But this principle remains valid even if – in order to bring theory closer to reality – the four simplificatory conditions are jettisoned one by one and the circulatory process restored to its complex reality.

Up to now I have supposed that the circulating medium is constituted by a good that does not have its own utility, but in fact this actually happens only very rarely – in the case of paper money. Hence all the points I have made up to now can be considered to refer to the typical case of paper money. And here my conclusions agree as we have seen with those of the quantity theory. Normally however money is constituted by a good that does have its own utility, for in general one single good serves for monetary and industrial uses. This means that it has two kinds of utility; a utility that depends on its quality as a good, and an instrumental utility deriving from its use as money. The former kind of utility varies in relation to this good's own curve of utility. The latter is calculated in the manner indicated, including among the circulating goods which together determine it that part of the good-money that is for industrial use. The quantity of this good available overall is distributed between the two uses. But just as the two fields for the two uses are linked, so the good-money will be distributed between them in such a way as to procure the same marginal level of utility in both, and the quantity of money that will enter into circulation will be such as to meet this equilibrium condition. But if the marginal utility of money as a good and its utility as an instrumental good coincide, it follows that to express and measure the value of money either may be used. Hence in the case of a circulating medium having its own utility too the principle of instrumental utility I have expounded has the same value; it coincides or is in full agreement with the principle that locates the value of money in the marginal utility of the good-money.

If we accept the cost of production principle as a hypothesis for money too, this means that production of the good-money is regulated by cost, and in this case equilibrium is established in correspondence with a level of production of this good such that at it the two marginal levels of utility are equal to each other and equal to the cost. In this case the instrumental utility is equal to the cost, and either may be employed equally to express and measure the value of money. Hence in this case the principle of the instrumental utility of money coincides and is in full agreement with the principle of the cost of production.

If in order to bring theory even closer to reality we suppose that in the unit of time the velocity of circulation of money is greater than one, then to find the marginal utility of money it is sufficient to replace the term $M$ as denominator of the expression representing this utility with the term $MV$, where $V$ represents the velocity of circulation of $M$. If we accept that the circulating medium is constituted both by metallic money and by credit money, then the marginal utility of the overall circulating medium is expressed by the original formula in this case too, provided that in the denominator the term $M$ is replaced by the term $M \cdot V + M_1 \cdot V_1$; where $M_1$ and $V_1$ represent the quantity of credit money and its velocity of circulation.

So then however complicated the circulatory process and whatever view we take of it the principle of the marginal utility of money I have expounded remains valid, and its ability to meet all possible cases confers on it a maximum level of generality. If we consider the case of a circulation purely of paper or a metallic circulation where for simplicity's sake we ignore the production and industrial use of precious metals, the principle expounded here harmonises in its conclusions with the quantity principle. If, taking account of the industrial usage of the good-money, we consider the latter in its effects on monetary circulation, the principle set out here harmonises with the principle of the marginal utility of the metal-good. Finally if we extend the laws of production equilibrium to precious metals too and treat their production as a function of their cost, the principle expounded here agrees in its conclusions with the cost of production principle and can therefore replace the latter. So the principle put forward in the text and illustrated perhaps too briefly in this note emerges as the general principle of the value of money that embraces, interpenetrates, and brings into harmony all the other unilateral principles; it replaces the various formulae corresponding to the various types of money with a single unique formula valid at the same time for paper money and metallic money, a formula that in its extreme generality brings the law of the value of any species of money back to general scientific principles.

But this acceptance of the theory under discussion does not mean a complete rejection of the quantity theory, of which I have made very broad application in another work (Fanno, 1913). The theory I have outlined is very far from excluding the quantity theory and in a sense serves to integrate that theory. While in fact the quantity theory provides the laws of the variations in the value of money or in the level of prices, the theory currently under discussion here provides the laws of variations in the general level of prices and in addition the laws of variations in individual prices that together determine that level. Therefore where study is limited to variations in the general level of prices it does not matter whether we state the problem on the basis of one or of the other principle, although the quantity principle will be preferable because it is simpler and easier to understand even though it is less rigorous. But where the task is to study variations in individual prices as well – as is the case with the problems under discussion in the text – only the other principle can provide an adequate solution.

10. The function involved here expresses the production cost not in terms of money, but rather in terms of units of sacrifice or in terms of the Cairnesian meaning of cost. Cost understood in this way is translated into monetary cost by dividing the value of the function expressing the marginal sacrifice by the marginal utility of money.

11. The terms of these equations represent in the first order the quota parts of the individual prices of the two joint goods that together form the global price and in the second order their global marginal cost. The formula of the prices is found by dividing the value of the function of the marginal utility of each good by the marginal utility of money. Now as we saw earlier, the marginal utility of money is represented by a fractional formula. Therefore the prices of the various goods are given by the marginal utility of each good divided by the fraction representing the marginal utility of money or more simply multiplied by the inverse

of this fraction. These complex operations are what produce the complex expressions of the prices indicated above. In a similar way the global monetary cost is obtained as a function of the psychological cost. (On the concept of psychological cost, see Cairnes, 1874, pp. 109 ff.)

12. Strictly speaking, as we have supposed for simplicity's sake in this case that in the two countries money is constituted by a good without any direct utility, this money should be considered as non-exportable and the export of it proposed in the text as impossible. But in order not to complicate this study inordinately I will assume it to be exportable and suppose that the transfer of this money from country *S* to country *T* is agreed and made possible owing to special monetary conventions between the two countries.
13. For a more thoroughgoing enumeration and specification of the conditions necessary for the attainment of international economic equilibrium, see Appendix III. International equilibrium is conditional on the equality of the prices of the two goods in the two countries. At the moment when this equality is reached, exchange of products and money ceases completely because equality of prices resolves into equality of comparative respective values of *A*, *B* and *M* in the two countries. As we know, when the comparative values are equal international exchange is impossible (Loria, 1901, pp. 85–9; see also Fanno, 1906, p. 218.)
14. See Chapter 3, Part I, s. 3 above.
15. Here I continue to speak of curves of demand and supply, despite the presence of several variables in the expressions employed to represent demand for and supply of the various goods, which prevents us from supplying graphs of this demand and supply in the shape of curves or at least of plane curves as generally employed. But for the sake of simplicity and brevity I will carry on speaking in terms of curves even though it is not really appropriate to do so.
16. Strictly speaking the overall increase in production or the sum of increases in *S* and *T* is smaller in this case than in the preceding case. Hence the amount by which the increase in production falls in country *S* is greater than the amount by which it increases in country *T*. But once these corrections have been made in the interests of accuracy, the final conclusion remains unchanged – that in this case the rise in production in *S* is smaller and in *T* greater than in the preceding case.
17. Ricardo (1982), i, p. 132; Mill (1909), BK III, ch. xxi, pp. 623–4; Bastable (1900), p. 185.
18. See Chapter 3, Part II, 5 above.
19. Ricardo (1982), i, p. 132; Mill (1909), BK III, ch. xxi, p. 850.
20. There is just one case where this is not necessary – where the marginal cost in country *T* is equal to or higher than the monopoly price in country *S*. In this case the producer in country *S* can sell in *T* at a price that is equal to or higher than the price on the domestic market, but this is quite obviously exceptional and as a general rule, given the existence of a monopoly in *S* and of free competition in *T*, the assumption is that prices are higher in *S* than in *T*.

21. Jannaccone (1913), pp. 329–33.
22. In this connection it seems appropriate to point out a singular fact deriving from the operation of supply at joint cost, namely this: sales by the country *S* producer, initially limited just to good *A*, may eventually also be extended to good *B*, even though the said producer does nothing to push the sale of this good in country *T*. Let us follow through the whole chain of actions and reactions to its remote effects deriving from the sale of good *A* below cost. We have seen that producers in *T*, who are obliged to sell good *A* at a price below the initial one eventually end up selling the two products *A* and *B* at a global price lower than the global cost. But the consequence is that as global production in *T* contracts because of the higher costs production of *A* and *B* falls and the price of *B* will tend to rise. Hence if to begin with the country *S* producer supplied good *B* in *T* at a price a little above the prevalent one and did not succeed in making any sales, he will now find himself in the position of being able to sell good *B* without lowering its original supply price. Therefore as a result of the operation of supply at joint cost dumping, which initially only affected good *A*, eventually extends automatically to good *B* as well.
23. Jannaccone (1913), pp. 250 ff.
24. For there to be exports of both goods it is sufficient that just one of them be offered at a price infinitesimally lower than before, in which case the global price falls below the cost in *T*. As global production in *T* falls the price of the other good will rise and the producer will find himself in a position to export that too but at a price equal to or greater than before.
25. This means that production for export with differentiation in pricing between the domestic and the foreign market is also possible and profitable in cases of uniform and growing costs. Here our conclusions chime with those of Professor Jannaccone (1914, pp. 267 ff.) but are at odds with those of Professor Cabiati (1914, p. 206) who erroneously supposes that such price discrimination can apply only to goods at falling costs.
26. Marshall (1920), Bk V, ch. vi, s. 5, p. 390.
27. From all this we may deduce that the elasticity of the sale curve is always greater than the elasticity of the corresponding demand curve. There is only one case where it is not greater, and that is where the good involved cannot be produced in the country concerned, in which case the sale curve coincides with the demand curve and has the same elasticity.
28. But in the long term the other good too can definitely be exported in given circumstances; this is because export of the first good depresses its price in *T* and brings about a decrease in production there with a consequent rise in the price of the other good. Now it can happen that as a result of this rise production for export, having previously been impossible now becomes advantageous. If so there will be production for export and both goods will be exported.
29. It would be more precise to call the regime under which coke and in general good *B* are sold a *duopoly* rather than a genuinely competitive

regime. Edgeworth first and then Pareto demonstrated how under a duopoly there is no stable equilibrium position, which means the problem is overdetermined and so insoluble (Edgeworth, 1925, i, pp. 116 ff.; Pareto, 1971, Appendices). In the present case however conditions are somewhat different. If one of the two goods is sold by each producer under a *duopoly* regime the other continues to be sold under a monopoly regime, so the problem is rather different, and its solution is therefore different as well.

30. See Chapter 2 Part II, s. 5.
31. See above Chapter 2 Part II, s. 2.
32. 'Suppose the same process to yield two joint products, one of which is controlled monopolistically but the other is not. Then, as shown above, if entry to the industry can be restricted, simple monopoly will make the output of both products less than it would be under simple competition. The whole of the non-monopolised joint product that is produced will be sold, but a part of the other product will probably be thrown away' (Pigou, 1912, p. 198 n. 1).

# Bibliography


Aupetit, A. (1901) *Essai sur la théorie générale de la Monnaie* (Paris).

Auspitz, R. and Lieben, R. (1889), *Untersuchungen über die Theorie des Preises* (Leipzig).

Bagiotti, T. (ed.) (1966a) *Essays in Honour of Marco Fanno* (Padua: Cedam).

Bagiotti, T. (1966b) 'Marco Fanno. A Neoclassical Curriculum', in T. Bagiotti (ed.), *Essays in Honour of Marco Fanno*, vol. II (Padua: Cedam).

Barone, E. (1894a) 'A proposito delle indagini del Fisher', *Giornale degli Economisti* (May).

Barone, E. (1894b) 'Sulla "consumers' rent"', *Giornale degli Economisti* (September).

Barone, E. (1894c) 'Sul trattamento di questioni dinamiche', *Giornale degli Economisti* (November).

Barone, E. (1896) 'Studi Sulla distribuzione', *Giornale degli Economisti* (February).

Barone (1908) *Principii di Economia Politica*.

Barone, E. (1912) 'Studi di economia finanziaria', *Giornale degli Economisti*, (April–August).

Bartoli, H. (1981) 'Marco Fanno e Achille Loria', *Rivista Internazionale di Scienze Economiche e Commerciali*.

Bastable, C.F. (1900) *The Theory of International Trade* (3rd edn) (London).

Becker, G.S. (1965) 'A Theory of Allocation of Time', *Economic Journal*, 75.

Bidard, C. (1986), 'Is von Neumann Square?', *Zeitschrift für Nationalökonomie*.

Böhm-Bawerk, E. von (1891) *The Positive Theory of Capital* (London).

Böhm-Bawerk, E. von (1902) *Histoire critique des théories de l'intérêt du Capital*, 12 vols (Paris).

Cabiati, A. (1914) 'Prime linee per una teoria del dumping', *Riforma Sociale* (March).

Cairnes, J.E. (1874) *Some Leading Principles of Political Economy* (London).

Clark, J.B. (1907) *Essentials of Economic Theory* (New York).

Clark, J.B. (1908) *The Distribution of Wealth* (New York).

Cournot, A. (1887) *Researches into the Mathematical Principle of the Theory of Wealth*, trans N.T. Bacon (New York).

Cunynghame, H.H.S. (1892) 'Some Improvements in Simple Geometric Methods etc.' *Economic Journal*, II.

Edgeworth, F.Y. (1881) *Mathematical Psychics* (London).

Edgeworth, F.Y. (1898) 'Professor Graziani on the Mathematical Theory of Monopoly,' *Economic Journal*, 8.

Edgeworth, F.Y. (1910a) 'Joint production', *Palgrave's Dictionary of Political Economy*, 2nd edn, vol. ii, (London).

Edgeworth, F.Y. (1910b) 'Elasticity', *Palgrave's Dictionary of Political Economy*, 2nd edn., vol. i (London).

Edgeworth, F.Y. (1910c) 'Applications of Probabilities to Economics', *Economic Journal* (June).
Edgeworth, F.Y. (1913) 'Contributions to the Theory of Railway Rates', *Economic Journal*, (June).
Edgeworth, F.Y. (1915) 'Recent Contributions to Mathematical Economics', *Economic Journal*, 15.
Edgeworth, F.Y. (1925) 'The Pure Theory of Monopoly', *Papers Relating to Political Economy*, 3 vols (London).
Fanno, M. (1906) *Espansione commerciale e coloniale degli Stati Moderni* (Turin).
Fanno, M. (1913) *Le Banche e il Mercato Monetario* (Rome).
Fanno, M. (1914) *Contributo alla teoria dell'offerta a costi congiunti*, Supplement to the *Giornale degli Economisti e Rivista di Statistica*.
Fanno, M. (1925) 'Alfredo Marshall', *Annali di Economia*, 2.
Fanno, M. (1926) *Contributo alla teoria economica dei beni succedanei* (Milano)
Fanno, M. (1932) 'Correlazione: tra prezze e curve statistiche di domanda e offerta', *Rivista Italiana di Statistica, Economia e Finanza*, 4.
Fanno, M. (1934) 'Della identità di alcuni teoremi di economia pura', *Giornale degli Economisti*, 74.
Ferrara, F. (1856) 'Introduzione' to vol. XI, serie 1, of *Bibliografica dell' Economista* (Turin).
Finoia, M. (1980), *Il Pensiero Economica Italiano 1859/1950* (Bologna : Cappelli).
Fisher, I. (1892) 'Mathematical Investigations in the Theory of Value and Prices', *Transactions of the Connecticut Academy*, vol. ix, part 1 (New Haven).
Fisher, I. (1907) *The Rate of Interest* (New York).
Fisher, I. (1912) *Elementary Principles of Economics*, 3rd edn (New York).
Flux, A.W. (1904) *Economic Principles* (London).
Goodwin, R.M. and Punzo, L.F. (1987), *The Dynamics of a Capitalist Economy. A Multi-sectoral Approach* (Boulder: Westview Press; Cambridge: Polity Press).
Graziani, A. (1908) *Istituzioni di Economia Politica* (Turin).
Halevi, J. (1987), 'Fanno, Marco', in J. Eatwell, M. Milgate and P. Newman, *The New Palgrave* (London: Macmillan).
Hicks, J.R. (1939) *Value and Capital* (Oxford: Clarendon Press).
Hicks, J.R. (1958) *A Revision of Demand Theory* (Oxford: Clarendon Press).
Jannaccone, P. (1901) *Il costo di produzione* (Turin).
Jannaccone, P. (1902) 'Questioni controverse nella teoria della traslazione delle imposte in regime di monopolio', *Riforma Sociale*.
Jannaccone, P. (1913) 'Note in margine alla replica del prof. Attilio Cabiati', *Riforma Sociale* (March).
Jannaccone, P. (1914) 'Il Dumping e la discriminazione dei prezzi' *Riforma Sociale* (March).
Jevons, W.S. (1931) *The Theory of Political Economy*, 4th edn (London).
Kinley, D. (1911) *Money* (New York).
Kurz, H. (1986) 'Classical and Early Neoclassical Economists on Joint Production', *Metroeconomica*, 37.

Lancaster, K. (1966) 'A New Approach to Consumer Theory', *Journal of Political Economy*, 74.

Landry, A. (1904) *L'intérêt du capital* (Paris).

Legris, A. (1994), 'L' "Hétérodoxie" de Marco Fanno et la Culture Économique Italienne de l'Entre-deux-Guerres', *Economies et Sociétés*, Serie Economia, *Historie de la Pensée Économique*, 19.

Loria, A. (1901) 'Notes on the theory of international trade', *Economic Journal*, (March).

Manfredini Gasparetto, M.L. (1992) *Marco Fanno. L'Uomo e l'Economista* (Padua: Cedam).

Marshall, A. (1907) *Principles of Economics*, 5th edn (London: Macmillan).

Marshall, A. (1920) *Principles of Economics*, 8th edn, Bk V (London).

Menger, C. (1976) *Principles of Economics*, trans J. Dingwall and B.F. Hoselitz (New York and London).

Mill, J.S. (1909) *Principles of Political Economy*, new edn (London); reprinted Fairfield, NJ (1987).

Morishima, M. (1952) 'On the Laws of Change of the Price System in an Economy which Contains Complementary Commodities', *Osaka Economic Papers*, I.

Morishima, M. (1959) 'The Problem of Intrinsic Complementarity and Separability of Goods', *Metroeconomica*, II.

Morishima, M. *et al.* (1973) *Theory of Demand, Real and Monetary* (Oxford: Clarendon Press).

Morishima, M. (1996) 'Morishima on Ricardo: Two Replies', *Cambridge Journal of Economics*, 20.

Mosak, J.L. (1944) *General Equilibrium Theory in International Trade* (Bloomington, Indiana: The Principia Press).

Moulin, H. (1995) 'On Additive Methods to Share Joint Costs', *The Japanese Economic Review*, 46.

Neumann, J.von (1945–6) 'A Model of General Economic Equilibrium', *The Review of Economic Studies*, 13.

Nicholson, J.S. (1897) *Principles of Political Economy*, 3 vols (London).

Pantaleoni, M. (1909), *Alcune osservazioni sui sindacati e sulle leghe* in *Scritti varii di Economia*, Serie II (Palermo).

Pantaleoni, M. (1910) 'L'identità della pressione teorica di qualunque imposta a parità di ammontare e la sua semiotica', *Giornale degli Economisti* (March).

Pantaleoni, M. (1957) *Pure Economics*, trans T. Boston Bruce (New York).

Pareto, V. (1892) 'La teoria dei prezzi dei sigg. Auspitz e Lieben', *Giornale degli Economisti* (March).

Pareto, V. (1893) 'Considerazioni sui principii fondamentali dell' economia politica pura', *Giornale degli Economisti* (October)

Pareto, V. (1896–7) *Cours d'économie politique*, 2 vols (Lausanne).

Pareto, V. (1971) *Manual of Political Economy*, trans A.S. Schwier, ed. A. S. Schwier and A.N. Page (London).

Pigou, A.C. (1903) 'Some Remarks on Utility', *Economic Journal* XIII.

Pigou, A.C. (1910) 'A Method of Determining the Numerical Value of Elasticities of Demand', *Economic Journal* (December).

Pigou, A.C. (1912) *Wealth and Welfare* (London).

Pigou, A.C. (1913a) 'Railway Rates and Joint Costs', *The Quarterly Journal of Economics* (August).

Pigou, A.C. (1913b) 'The interdependence of different sources of demand and supply', *Economic Journal*.

Punzo, L.F. and Velupillai, K. (1984), 'Multisector Models and Joint Production', in F. van der Ploek (ed.), *Mathematical Models in Economics*, vol III of *Handbook of Applicable Mathematics* (London: John Wiley).

Rae, J. (1834) *The Statement of Some New Principles on the Subject of Political Economy Exposing the Fallacies of the System of Free Trade and of Some other Doctrines Maintained in the Wealth of Nations* (Boston).

Ricardo, D. (1982) *On the Principles of Political Economy and Taxation*, in P. Sraffa and M.H. Dobb (eds), *The Works and Correspondence of David Ricardo*, 11 vols (Cambridge).

Ricci, U. (1906), 'Curve piane di offerta dei prodotti', *Giornale degli Economisti*, (September).

Salvadori, N. and Steedman, J. (1988) 'Joint Production Analysis in a Sraffian Framework', *Bulletin of Economic Research* 98.

Schefold, B. (1980) 'Von Neumann and Sraffa: Mathematical Equivalence and Conceptual Difference', *Economic Journal* 90.

Schefold, B. (1984), 'Sraffa and Applied Economics: Joint Production', *Political Economy* 1.

Schefold, B. (1987), 'Joint Production in Linear Models', in J. Eatwell, M. Milgate and P. Newman, *The New Palgrave* (London: Macmillan).

Shoup, C.S. (1966) 'Public Goods and Joint Production', in T. Bagiotti (ed.), *Essays in Honour of Marco Fanno* (Padua: Cedam).

Schumpeter, J.A. (1908) *Das Wesen und der Hauptinhalt der theoretischen Nationalökonomie* (Leipzig).

Schumpeter, J.A. (1954), *History of Economic Analysis* (New York: Oxford University Press).

Sidgwick, H. (1887), *The Principles of Political Economy* (London).

Smart, W. (1912) *The Distribution of Income* (London).

Sraffa, P. (1960), *Production of Commodities by Means of Commodities*, (Cambridge: Cambridge University Press).

Stackelberg, H. von (1934) *Marktform und Gleichgewicht* (Berlin: Verlag von Julius Springer).

Steedman, I. (1984) 'L'Importance Empirique de la Production Jointe', in C. Bidard, *La Production Jointe, Noveaux Débats* (Paris; Economica).

Taussig, F.W. (1911) *Principles of Economics*, 2 vols (New York).

Taussig, F.W. (1913) 'The Tariff Act of 1913', *Quarterly Journal of Economics*, 8.

Taussig, F.W. (1914) 'Abraham Lincoln on the Tariff: A Myth', *Quarterly Journal of Economics*, 8.

Valenti, G. (1909) *Principii di scienza economica* (Florence).

Vecchio, G. del (1909) 'I principii della teoria economica della moneta', *Giornale degli Economisti*, xxxix, II.

Walras, L. (1954) *Elements of Pure Economics*, trans W. Jaffé (London).

Wicksell, K. (1934) *Lectures on Political Economy*, English edn (London: Routledge & Sons).

# Index